Engineering Skills for Career Success

Charles K. Alexander
Cleveland State University

James A. Watson
Watson Associates

ENGINEERING SKILLS FOR CAREER SUCCESS
International Edition 2014

Exclusive rights by McGraw-Hill Education (Asia), for manufacture and export. This book cannot be re-exported from the country to which it is sold by McGraw-Hill. This International Edition is not to be sold or purchased in North America and contains content that is different from its North American version.

Published by McGraw-Hill Education, 2 Penn Plaza, New York, NY 10121. Copyright © 2014 by McGraw-Hill Education. All rights reserved. No part of this publication may be reproduced or distributed in any form or by any means, or stored in a database or retrieval system, without the prior written consent of publisher, including, but not limited to, in any network or other electronic storage or transmission, or broadcast for distance learning.
Some ancillaries, including electronic and print components, may not be available to customers outside the United States

10 09 08 07 06 05 04 03 02 01
20 15 14
CTP SLP

All credits appearing on page or at the end of the book are considered to be an extension of the copyright page.

When ordering this title, use ISBN 978-981-4581-89-9 or MHID 981-4581-89-5

Printed in Singapore

www.mhhe.com

ABOUT THE AUTHORS

Charles K. Alexander, PhD, HDr.Eng., LFIEEE, PE, is a professor of electrical and computer engineering in the Fenn College of Engineering's Electrical and Computer Engineering Department at Cleveland State University. He is also the director of the Center for Research in Electronics and Aerospace Technology (CREATE).

From 2002 until 2006, he was dean of the Fenn College of Engineering and from 1998 until 2002 he was interim director (2000 and 2001) of the Institute for Corrosion and Multiphase Technologies and Stocker Visiting Professor of electrical engineering and computer science at Ohio University. From 1994 to 1996 he was dean of the College of Engineering and Computer Science at California State University, Northridge. From 1989 to 1994 he was acting dean of the College of Engineering at Temple University, and from 1986 to 1989 he was a professor and chairman of the Department of Electrical Engineering at Temple. From 1980 to 1986 he held the same positions at Tennessee Technological University. He was an associate professor and a professor of electrical engineering at Youngstown State University from 1972 to 1980, where he was named Distinguished Professor in 1977 in recognition of "outstanding teaching and research." He was assistant professor of electrical engineering at Ohio University in 1971 and 1972. He received the Ph.D. (1971) and M.S.E.E. (1967) from Ohio University and B.S.E.E. (1965) from Ohio Northern University.

Dr. Alexander has been a consultant to twenty-three companies and governmental organizations, including the Air Force and Navy and several law firms. He has received over $100 million (in 2012 dollars) in research and development funds for projects ranging from solar energy to software engineering. He has authored 40 publications, including a workbook and a videotape lecture series, and is coauthor of *Fundamentals of Electric Circuits*(1st, 2nd, 3rd, 4th, and 5th editions), *Applied Circuit Analysis, Problem Solving Made* Almost *Easy*, and the fifth edition of the *Standard Handbook of Electronic Engineering* with McGraw-Hill. He has made more than 500 paper, professional, and technical presentations.

Dr. Alexander is a life fellow of the IEEE and served as its president and CEO in 1997. In 1993 and 1994 he was IEEE vice president, professional activities, and chair of the United States Activities Board (USAB). In 1991 and 1992 he was Region 2 Director, serving on the Regional Activities Board (RAB) and USAB. He has also been a member of the Educational Activities Board. He served as chair of the USAB Member Activities Council and vice chair of the USAB Professional Activities Council for Engineers, and he chaired the RAB Student Activities Committee and the USAB Student Professional Awareness Committee.

He received the 2013 IEEE Undergraduate Teaching Award. In 2012 he received the Jim Watson Student Professional Awareness Achievement Award. In 2009, he received the honorary doctor of engineering from Ohio Northern University. In 1998 he received the Distinguished Engineering Education Achievement Award from the Engineering Council, and in 1996 he received the Distinguished Engineering Education Leadership Award from the same group. When he became a fellow of the IEEE in 1994, the citation read "for leadership in the field of engineering education and the professional development of engineering students." In 1984 he received the IEEE Centennial Medal, and in 1983 he received the IEEE/RAB Innovation Award, given to the IEEE member who best contributes to RAB's goals and objectives. He has received numerous awards not listed here.

James A. Watson, PE, President, Watson Associates received a bachelor's degree in electrical engineering from Purdue University, completed postgraduate engineering courses at Youngstown State University, and is a graduate of the Executive Management Program at the University of Michigan.

He is a Registered Professional Engineer in the State of Ohio, a life senior member of IEEE, and an international speaker for the IEEE Student Professional Awareness Program and IEEE Member Professional Awareness Program.

Following a 36-year career with Ohio Edison Company, an electric utility company in which he held numerous engineering and management positions, Jim took an early retirement to devote himself full time to his consulting activities. Founded in 1989, Watson Associates provides information, resources, and workshops for career management and other professional related topics.

He is the author of the WRITETALK© and ProSkills© programs, designed to incorporate communication and other career development skills and exercises within the university engineering technical curriculum. Watson Associates has directed the use of this program for more than 24,000 students at 14 major universities.

Jim initiated his career development work with engineering students in 1975 at Youngstown State University and the University of Akron. In 1982, as an Industrial Representative for the IEEE Student Activities Committee, he expanded his volunteer activities with students by becoming a national speaker for Student Professional Awareness Conferences (S-PACs).

Based on the early success of the S-PAC program, Jim chaired a committee in 1992 to formally structure a similar program to emphasize professional awareness for practicing professionals called Member Professional Awareness Conference (M-PAC).

Jim received numerous awards from IEEE in recognition of his volunteer activities including the USAB Citation of Honor (1987), IEEE USAB Professional Achievement

Award (1990), IEEE RAB Innovational Award (1990), IEEE Third Millennium Metal (2000), and IEEE-USA Citation of Honor (2001).

In 2011, Jim received the Robert S. Walleigh Distinguished Contributions to Engineering Professionalism Award for his more than 40 years of leadership in promoting the professional development of IEEE members. This is IEEE-USA's highest recognition for volunteer activities.

In 2011, IEEE-USA established the "IEEE-USA Jim Watson Student Professional Awareness Award" to recognize key individuals for their contributions to Student Professional Awareness Achievement Activities in the United States.

Jim has published articles on career development in *IEEE Antennas & Propagation, IEEE Potentials, IEEE Women in Engineering, HKN Bridge*, and *Wireless Systems Design*.

As an international speaker, Jim has delivered more than 2,000 presentations in the United States, Canada, Europe, and Asia to a total audience of over 100,000, including 55,000 students at 157 universities.

CONTENTS

PREFACE

Rarely does anyone have the opportunity to be able to convert forty years (together over eighty people years) into a book. We are honored to write this book and combine our more than eighty years of research on what is involved in becoming a successful engineer. In addition, we have developed lectures and teaching modules (WRITETALK© and ProSkills©) to help prospective engineers and working engineers develop and enhance the skills needed in order to be successful.

As colleagues who have collaborated and worked on various projects for over forty years, we are grateful for this opportunity to share our knowledge and experiences with you. Some of the results of our research were originally presented as chapter openers for several of the chapters in Alexander and Sadiku's textbook, *Fundamentals of Electric Circuits*, published by McGraw-Hill. Many users of that book suggested that we write a book dealing exclusively with these topics. We are pleased to offer *Engineering Skills for Career Success* in response to those suggestions.

We are firmly convinced that if the principals that are contained herein are followed, you will indeed have a more successful career. In addition, you will also have a great deal more fun!

We would like to give a very personal thanks to Judith Jurgens Watson for all that she has done to make this book a success. Her professional approach to written communications, her university background as a communications instructor, along with her organizational skills and unique perspective contributed greatly to the quality of this book. Thank you Judy!

We would also like to express our appreciation for the support and encouragement we have received from Hannah Zhang Alexander. Her excitement, optimism, and just a general belief in this project have been wonderful! Thank you Hannah!

In addition, there have been several individuals from McGraw-Hill who not only have made this project much better but also have been responsible for making it possible. We thank the following: Brand Manager Raghu Srinivasan, Executive Editor Bill Stenquist, Developmental Editor Katie Neubauer, Executive Marketing Manager Curt Reynolds, and Project Manager Lisa Bruflodt.

This book is based on the soft skills that are needed by the successful engineer. The book is organized into the following three parts:

SECTION I—BASIC SKILLS

The focus of Chapters 1 through 6 is on basic skills engineers need to have in order to be successful. Our discussion emphasizes how to develop and enhance these basic skills.

SECTION II—ENGINEERING SKILLS

Chapters 7 through 12 focus on further refining and developing the skills established in Section I. In addition, material is presented to enable the student to make sound career decisions. Along with this, students are exposed to the importance of documenting activities, developing team building skills, and learning how to handle a wide variety of ethical dilemmas.

SECTION III—CAREER SKILLS

In Chapters 13 through 17, the student is exposed to advanced assets designed to continue the development of a successful career. A variety of activities that can enhance one's career are presented in detail. These include being able to apply the principles of systems engineering to problem solving, design, and career planning; understanding the elements of project management; developing personal networks; benefitting from co-op and internship experiences; and understanding the importance of a graduate education.

In Chapter 1, "Four Principles for Earning Higher Grades," you will learn how to maximize your learning experiences in the classroom and how to enhance your potential to have an exciting career. You will learn how to start the process to be successful in your career. You will learn the most important tools to maximize your learning experiences in the classroom. The most significant of these tools are, in priority order, asking questions; studying in study teams; joining and being active in student organizations; and taking advantage of outside-the-classroom human resources, such as faculty during office hours and tutors. These are the four things that always lead to higher grades and enhanced learning. The most important lesson is that you need to be actively involved in your educational process.

In Chapter 2, "Develop Your Skill Set," you will learn how to specifically develop and enhance those skills ABET (the organization that academically accredits all engineering programs in the United States) feels are most important to a successful engineering career. ABET has classified these skills into eleven general areas. You will learn techniques for developing and enhancing each of these skills.

Chapter 3, "Time Management Skills," discusses the most important resource of time. It uses engineering tools and the basic principles of managing time as a foundation to show you how to customize individual time management systems. This chapter demonstrates how a time management process can be easy to implement and use and why time management is important to your career success. An exercise is included to show you how to develop a plan designed to meet individual work habits.

Chapter 4, "Communication Skills," is the first of two chapters on communication. The content shows you how career success is proportional to the ability to communicate through reading, writing, listening, and speaking. The chapter focus is on reading and writing effectively as an engineer, and includes several exercises to enhance these skills.

Chapter 5, "Professional Presentation Skills," continues the discussion on communication with the focus on public speaking. In addition to reviewing the principles of public speaking, this chapter includes exercises to demonstrate how engineering concepts can be used to enhance presentation skills. Resources include skill evaluations to show you specific methods to improve speaking skills and to gain self-confidence to reduce stress.

Chapter 6, "Information Capturing Skills," identifies an easy and effective way for you to capture information relating to activities during the academic experience. The chapter focus is on how information capturing is the foundation for building a personal portfolio, and how portfolios are a valuable resource to develop effective resumes and prepare for job interviews. The skills you develop in the exercises in this chapter are also essential in future working careers.

Chapter 7, "Engineering Disciplines—A Fantastic Way to Have Fun!," is designed to help you better understand the broad field of engineering. An argument for the five major drivers for engineering jobs in the first part of the twenty-first century is detailed in a way that will help you make career decisions. These are "manufacturing"; "affordable, high-quality healthcare"; "revitalizing and enhancing the infrastructure"; "energy independence and economic stability"; and "enhancing the environment." Through surfing engineering society websites, you will learn the breadth and depth of engineering disciplines and subdisciplines.

Chapter 8, "The Skill to Apply the Systems Approach to Engineering to Enhance Your Career," focuses on the elements of systems engineering and how that can help you to plan and enhance your career. What you learn in this chapter will help you to be more organized in your approach not only to your career but to engineering design, research, and problem solving. The main elements that will be covered are defining the problem or task, decomposing the problem of a task, defining the elements of the solution process, setting goals, identifying constraints, selecting alternatives, evaluation and trade-offs, determining outcomes and trade-offs, and brainstorming.

Chapter 9, "Project Management and How to Use It as a Skill to Enhance Career Success," present the elements of project management and how those elements can help you become a better engineer. In engineering, you are hired for what you can do and not for what you know. In addition, you will always work in groups and on projects. The more you know about project management the better you can perform your tasks. This chapter describes the entire project management process. The most important elements of this process are planning; communicating; organizing; motivating; directing; controlling; and finally, monitoring, evaluating, and revising all the elements. You will also learn how to take advantage of opportunities to enhance project management skills and the impact of project management skills on career success.

Chapter 10, "Team Building Skills," starts with a discussion of the steps used in building successful teams. It includes a self-evaluation exercise to help you identify various personality traits. Additional information shows how personality traits can be applied most effectively to team situations. This chapter emphasizes why learning how to work on a team will greatly enhance your working career success.

Chapter 11, "Engineering Skills for Ethical Dilemmas," demonstrates how to recognize and handle ethical situations. Exercises show how engineering tools and techniques can be used to solve ethical dilemmas. Because most engineers encounter ethical challenges in their working career, this chapter prepares you to effectively handle ethical situations.

Chapter 12, "Leadership Skills," outlines why learning how to be a leader and how to work with leaders is important to career success. Although most engineers are not born leaders, this is a skill that can be developed by a systematic process. This process starts by learning how to control personal time and activities, and then by expanding the application of these skills to leading others. This chapter will provide information and exercises for you to participate in leadership opportunities in school, in the professional environment, and in community activities. These experiences will prepare you to work in group situations as a leader and under the leadership of others.

Chapter 13, "Career Management Skills," focuses on how to plan and control your career. This process includes establishing personal and professional visions, identifying a strategy of goals and skill needs, developing and implementing an action plan, and documenting and evaluating results. Because career planning should start at the beginning of the academic program, the chapter discusses how you can maximize the value of all courses by relating content to career plans. An initial career plan is developed in exercises in this chapter, and you are encouraged to keep this career plan dynamic and updated as needed.

Chapter 14, "Networking and Professional Development Skills," emphasizes the importance of personal networking and professional development. Topics include how to make positive impressions when meeting people for the first time; the importance of active listening, and how to take advantage of experiences in classes, laboratories, and study groups to develop effective networking skills. Exercises show you how to build networks by active involvement in professional engineering organizations. In addition to obtaining technical information by participation in professional organizations, you can also increase your job opportunities, promotions, and career success.

Chapter 15, "Co-Op/Internship and Other Practical Experiences and How This Can Enhance Career Success," presents the advantages of having such experiences and how to obtain the most out of such experiences. Engineering is a practice oriented profession and a co-op/internship experience can significantly contribute to your success. You will also learn how to have a good co-op experience in a school without a co-op program.

In Chapter 16, "Graduate School, Maybe the Ultimate Way to Enhance Career Success!" you will learn whether you need to go to graduate school and if you do, where to go and which degrees you should pursue. You learn about a bachelor's degree, a master's degree, and a doctoral degree. You find clear guidance on whether or not you should go to graduate school and when you should go. The chapter will enable you to know whether or not you should pursue an MBA. You will find that there are many myths about going to graduate school making that decision process much easier.

Finally, in Chapter 17, "Through the Looking Glass, Bring All the Skills You Have Acquired to Build a Successful Career," explains how to transition from college into a successful working career. Part of this transition includes understanding the real world and how it is different from academic life. Information in this chapter is based on case studies and personal experiences in business/industry and academia.

Writing this book has been a joy of a lifetime (actually two lifetimes)! We sincerely wish that you will enjoy using this book to help you create a successful career as much as we enjoyed writing it!

James A. Watson **Charles K. Alexander**

We humbly dedicate this book to Judy and Hannah!

1 Four Principles for Earning Higher Grades

"But there are advantages to being elected President. The day after I was elected, I had my high school grades classified Top Secret."

—Ronald Reagan

1.1 Introduction

The title for this chapter could have been changed to "How to Earn Grades the Easy Way." Of course there really is no way to earn higher the easy way. You will be surprised, however, that if you follow the four principles we discuss in this chapter you will earn higher grades with a lot less effort!

When students approach professors and ask them how to improve their grades, the typical response (from the students themselves as well as the professor) is that the student needs to study more and to work more problems. While studying and working problems are very important, most students will not see any marked improvement by increasing these activities. In fact, most students do not really need increased studying and problem working. The best way for almost all students to improve their grades is to do the following, in priority order:

The best way to improve your grades:

1. ask questions in class,
2. work together in study teams,
3. join and be *active* in your professional society, and
4. take advantage of your people resources!

The important thing to realize is that none of these actions involve studying more and working more problems. Through many years of experience, we have found this method to work in every case! Another benefit is that this actually can make taking classes more fun! The method can also make classes more fun for teachers.

Before going into detail as to how best to accomplish each of the four activities, let us discuss, in a general way, each of the activities and why they work. Remember

LEARNING OBJECTIVES

By using the information and exercises in this chapter you will be able to:

- Develop an awareness of the importance of becoming *actively* involved in your educational program.
- Enhance your ability to actively participate in your educational process.
- Become aware of the importance of asking questions in class.
- Enhance your communication skills especially your listening and speaking skills.
- Develop skills to better understand the technical content of classes in which you are participating.
- Develop tools to help you achieve a more positive attitude toward lifelong learning.
- Develop a skill set that will generate, as a metric, higher grades.
- Build your teaming skills.
- Expand your active network.

we listed each of the principles in priority order. We discuss the importance of asking questions in class last, since this is the most important.

The second principle is **work together in study teams.** Engineering is actually a people profession, a theme we use throughout the book. Thus, the more we interact with other students the more we develop the important skills we will need as engineers. Forming a study team is truly a powerful learning tool. Students tend to learn better from people nearer their own age. So, asking a fellow student to explain something may mean that you will better understand the answer. Additionally, many professors have said they learn a topic better by teaching that topic. So, when a member of the study team asks another member of the study team a question, the student who responds will naturally learn the topic better.

The third principle is **join and be *active* in your student professional organization.** Care is needed here because some students can actually become too involved, hurting their grades rather than improving them. We discuss this fact later. The advantages of following this principle are manyfold, but the operative word is *active*! First, you are now in a group that includes upperclassmen, many of whom have already taken the classes you are taking. You now have a group that you can ask to explain material you are currently studying. Likewise, you will be asked to explain material to students currently enrolled in classes you have already taken. Indirectly, having a student organization that is doing well by having a large number of students actively involved tends to improve the esprit de corps of the academic department. This can lead to faculty members and fellow students who are more excited, creating an enhanced academic environment within the classroom. Naturally faculty members will, most likely, do a better job of teaching!

Our fourth principle, **take advantage of your people resources,** is, surprisingly, underutilized by a large number of students. The best people resource is the instructor.

Bring your professor thoughtful questions, during the professor's office hours. Some professors will also allow students to ask them technical questions related to classes they, personally, may not be teaching at the time. Finally, a lot of programs supply tutors for students who need them.

Now we come to the first principle, **ask questions in class.** For a number of years, we thought that asking questions in class was a powerful learning technique because students needed to formulate a question and then needed to interact with the professor answering the question. However, that did not completely explain why students did so much better by merely asking questions. Eventually, we really did understand why this works so well. When a student asks a question in class, they move from being just passively involved in the class to being actively involved in the educational process. This is clearly the most important principle because once you are actively involved in the educational process, you will better understand the material you are studying.

Asking questions in class is clearly the most important principle because once you are actively involved in the educational process, you will better understand the material you are studying.

Following our principles can actually lead to higher grades without additional time. And since it relies on activities outside the classical study approach, it also requires a different focus and skill sets (not just reading) and is much less boring! Although there are a lot of things at work here, the biggest difference is that now you are actively involved in the learning process.

Following these four principles can be seen as working smarter not harder!

1.2 Asking Questions in Class

Asking questions in class is the single most important thing you can do to enhance your learning experiences. Clearly, formulating a question and then asking it in class is very beneficial in itself. Responding to the instructor's response also leads to better understanding. But the most important part of this process is that now you are *actively involved* in the learning process.

Once a student is actively involved in the learning process, several significant things take place. The first is that the student is now in control. We all do better with less effort when we have a level of control over our learning or working environment—we are no longer passive participants in the process.

Asking questions in class forces you to become actively involved in the learning process and allows you to be in control of that process. Both of these phenomena are present in interactive video games.

An interesting personal anecdote is that if a student in one of Dr. Alexander's classes says, "I have sat in your class for six weeks and have not understood a word you have said!" Even though that is not really a question, and could infuriate some professors, that student can now be helped! Because the student is now actively involved, the professor can begin the process of finding out, with a series of simple questions, just how much the student understands and what the professor needs to tell the student. Generally, such students really do understand more than they think they do and can be brought up to speed fairly quickly. Which brings us to another important issue, we have found that no matter what the question, there is a large group of students in the class that benefit from the answer (you have all heard that there are no dumb questions).

Most of you have been repeatedly told during your academic careers that questions in class are always encouraged. Unfortunately, you may have learned that many teachers say this but, in truth, really do not mean it. The good news is that once you finally enter an engineering program, a large percentage of faculty members really do want you to ask questions in class. You can quickly learn which professors want questions and which do not.

Asking questions will take some patience on your part and you will need to work on the process for a while before you become comfortable. If you are nervous about just jumping in with a question in class, prepare a written question the night before and then read it in class.

When you have asked your question, you may not feel you fully understand the answer. Now you must follow up your question with a response. The worst thing you can say is, "I do not understand what you said." Most likely the professor will repeat what he or she said before. It is much better to rephrase the initial response from the professor in your own words. The best thing about this is that even if you are way off target, you have encouraged the professor to respond differently the second time, which should help your understanding. Again, you are now actively involved in the learning process.

Good luck in enhancing one of your most important skills as an engineer.

EXAMPLE 1.2.1

Ask a question in a math class as a learning aid. Develop a question to be asked in class and record the professor's response and your followup. Make sure there is interaction on the original question.

■ Solution

In a calculus class, a student did not fully understand part of a previous lecture where the professor had solved a calculus problem using integration by parts. Since the student did not remember how to integrate by parts, she asked the professor to explain what he meant by solving using integration by parts.

The professor responded, "Integration by parts means that you use the following formula, $\int u dv = uv - \int v du$." Now the poor student did not understand what the professor had said. The student, had two choices, she could state that she did not understand what the professor had said, which would probably have caused the professor to give the same response.

Or, the best approach, she tried to use her own words to give her understanding and cause the professor to modify his response to be more helpful.

She said, "Is this true for all functions, or is it only for certain classes of functions? Additionally, is there a simple proof for this relationship?"

In this case, this response did allow the professor to be more helpful to the student as well as the rest of the class.

PRACTICE PROBLEM 1.2.1

Pick a math class where you can develop a question that you will ask in that class. Record the question and the response from the professor.

EXAMPLE 1.2.2

Ask a question in an engineering class as a learning aid. Develop a question to be asked in class and record the professor's response and your followup. Make sure there is interaction on the original question.

■ **Solution**

For this example, we will use a question in a statics class, although other classes would be equally effective. The professor is talking about balancing forces acting on an object.

The first question the student asks is, "Why do the forces need to be balanced?" The professor responds, "The forces need to be balanced in order for the object on which the forces are acting does not move." The student then asks, "How do we know when the forces are balanced?"

The professor responds, "When we are working with forces in a two-demsional system, all we need to do is to sum all the forces as vectors and make sure that the sum equal zero."

At this point the student responds, "How do we know the forces cancel each other? And, can you give us an example of how this works?"

Asking for an example is always a good thing to do since the responder then must give an example with numbers in it. The professor responds, "First we convert all the forces into the Cartesian coordinate system. This will allow us to add up all the forces and see if the algebraic sum equals zero."

The professor then proceeds to give the following example, "Consider the object below with the indicated forces acting on it."

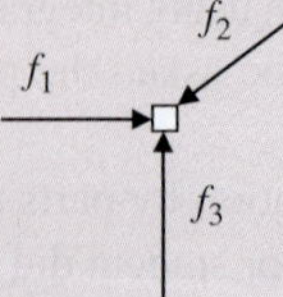

Figure 1.1
Example from a statics class.

The diagram in Figure 1.1 represents the following forces: $f_1 = 2\angle 0°$ newtons, $f_2 = 2\angle -135°$ newtons, and $f_3 = 3\angle 90°$ newtons. The first question: Is the object at rest? We can answer that by converting all of the forces into Cartesian values and see if they add up to zero.

$$f_1 = (2 + j0)\,\text{newtons},\ f_2 = (-1.4142 - j1.4142)\,\text{newtons, and } f_3 = j3\,\text{newtons}$$

Adding these together we get,

$$f_1 + f_2 + f_3 = 2 - 1.4142 + j(-1.4142 + 3) = (0.5858 + j1.5858)\,\text{newtons} \neq 0.$$

Since we have a resultant force, the object cannot be at rest and is accelerating in the direction of the resultant force. Now if we add a fourth force, $f_4 = (-0.5858 - j1.5858)$ newtons, we now have balanced the forces and the object can be at rest.

PRACTICE PROBLEM 1.2.2

Pick an engineering class where you can develop a question that you will ask in that class. Record the question and the response from the professor.

1.3 Forming Study Groups or Study Teams

You will notice that we strongly encourage **study teams** throughout the book (see especially the chapter on team building skills). This is the second most powerful technique for enhancing learning! What is most effective about this activity is that you will have to spend part of your time in the role of a teacher. Many will tell you that you can never really learn a subject until you teach it. Again, active participation is the most important part of this activity.

Even an hour a week can be quite productive, however, more time devoted to this activity will yield impressive results. You need a team that will allow you to have your questions answered adequately and that you can spend a significant part of your time answering other students' questions. Please do not worry if at first all you do is ask questions, time will help you develop the ability to answer questions. For this to really be effective, the members of the team need to be tolerant of responses by the team members. A study team cannot be effective if the members are afraid of being wrong.

How do you form a study team? There are basically two ways to become part of such a team. One is to be asked to join a team or ask to join an existing team. The other is for you to form a team. We are not sure whether the best study team is one made up of a range of students from the class (from individuals who are struggling to learn the material to those who seem to grasp the topics quickly) or a team formed of students with a like background and ability. Both of these have merits and it is possible that the best team for an individual will vary widely depending on the individual.

Making the most of this effort will require some effort from you. If you or another student asks a question, then your responsibility is to make sure you interact until you understand the answer and can put that answer into your own words. If you are not sure of a response, you need to consult other sources such as your textbook to help you understand that response. If your understanding of the response is different from the original response, please share your understanding with the study team the next time you meet or by e-mail. By the way, putting responses into words is also a very good way to learn material and to retain it as well as enhancing your writing skills.

A good task for study teams would be to assign members of the teams to give short summaries of lectures to the teams on a rotating basis. You might summarize Monday's lecture and another team member would be responsible for Wednesday's lecture. If you have three members, the next member would summarize Friday's lecture.

That brings us to another point; you need at least three members for a team. Having just two members in a study team places too much responsibility on each of the members and is clearly less effective. The important thing is that everyone must actively participate!

Having just two members in a study team places too much responsibility on each of the members and is clearly less effective. The important thing is that everyone must actively participate!

EXAMPLE 1.3.1

Select a topic from one of your math classes as a discussion topic for your study group. Assign one or more members the responsibility of making an initial presentation on the topic. Make sure there is a lot of discussion relative to the topic.

■ **Solution**

The student who initiated the question on integration by parts (Example 1.2.1) felt that there still was a lot to be learned about the topic and decided to present a discussion on the topic to her study group complete with a simple example. Here is her presentation.

There is a whole class of problems where integration by parts is an easier approach. We begin with

$$d[f(t)g(t)]/dt = \{[d[f(t)]/dt\}g(t) + f(t)[d[g(t)]/dt$$

Now all we need to do is to integrate both sides, which results in

$$f(t)g(t) = \int f'(t)g(t)\,dt + \int f(t)g'(t)\,dt$$

This then leads to the indefinite form:

$$\int f(t)g'(t)\,dt = f(t)g(t) - \int f'(t)g(t)\,dt$$

and the definite form:

$$\int_a^b f(t)g'(t)\,dt = \left[f(t)g(t)\right]_a^b - \int_a^b f'(t)g(t)\,dt$$

Sometimes the indefinite form is represented as

$$\int u\,dv = uv - \int v\,du$$

Let us try an example. Find the $\int_0^t \tau e^{-\tau} d\tau$. First we let $f(t) = t$, which gives $f'(t) = 1$ and then let $g'(t) = e^{-t}$, which gives $g(t) = -e^{-t}$. Now we can use Equation 1 to solve for the required integration.

$$\begin{aligned}\int_0^t \tau e^{-\tau} d\tau &= \left[\tau(-e^{-\tau})\right]_0^t - \int_0^t (-e^{-\tau}) d\tau \\ &= -te^{-t} - (0) - \left[e^{-\tau}\right]_0^t = -te^{-t} - e^{-t} + 1\end{aligned}$$

Does this solution make sense? If we differentiate the answer we obtain the original value to be integrated. Since we have a definite integral, the value of the integration going from 0 to 0+ must be equal to zero, this condition is also met. So, the answer is correct.

PRACTICE PROBLEM 1.3.1

Select a topic from one of your math classes as a discussion topic for your study group. Assign one or more members the responsibility of making an initial presentation on the topic. Make sure there is a lot of discussion relative to the topic.

1.4 Joining and Being Active in Your Professional Organizations

Joining and being active in your professional organizations (AIAA, AIChE, ASCE, ASHRAE, ASME, IEEE, NSBE, SAE, SWE, etc.) is the third most important thing you can do to enhance your learning environment. The key word is "active!" Clearly, being active in your society will help you enhance your skill set. The network you develop, however, can also be an important source of technical information. Generally, it is easier to learn from your peers than your professors especially when the professors are old enough to be parent figures. Asking a question of an upper-class student who had the class the year before can lead to valuable insights, especially if they had the same professor you have. Again, you can enhance your experience by becoming a teacher to the fellow student who can validate your understanding or challenge you. We cover this topic more completely in the chapter on networking and professional development skills.

We now will focus on a couple of exercises to help enhance your academic environment.

PRACTICE PROBLEM 1.4.1

Select a committee from your student organization to join. Once you have become actively involved with that committee, find a committee member who has had one or more of your engineering, math, or science courses, ones you are currently taking.

From one of those courses, choose a problem or topic about which you still need a better understanding. Ask the upperclassperson to help you with that topic or problem. Make sure that you are as specific as possible. Record your efforts and the results.

We ask you to record your activities and responses throughout the book! This is because you are just learning valuable skills to enhance learning. As you become more proficient, you will not need to do so much documentation later. However, until you learn to master the skill, the documentation part is extremely important!

PRACTICE PROBLEM 1.4.2

Using the committee you joined in Practice Problem 1.4.1, find an underclassperson who is currently taking a technical, math, or science course you have already had. If there is no such student currently on the committee, recruit one with those qualifications.

As part of your recruiting efforts, emphasize the value of active involvement in a student organization. Ask this student about one of the courses they are currently taking and try to help that student better understand what they are studying. Record your efforts and the response to them.

1.5 Taking Advantage of Outside-the-Classroom Resources

The fourth most important activity in which you can engage is **seek out help from others** such as the professor during their office hours. You would be surprised at how few students seek out their professors after class. You can go to your advisor for help. Sometimes you can approach another faculty member for help. Many schools supply free tutors for the engineering students. You can always pay for a tutor if necessary.

PRACTICE PROBLEM 1.5.1

Select a professor from a class in which you are currently enrolled. Formulate a good question from a lecture you have recently heard in which you felt you did not understand everything fully. During the professor's office hours, ask the professor to help you better understand the topic or problem.

It is quite helpful when you ask the question if you indicate that you have given the problem or topic some careful thought. Record you efforts and the professor's response.

PRACTICE PROBLEM 1.5.2

Analyze your weaknesses relative to one of your fundamental background areas. We suggest you look at your math skills and pick one of them such as algebra or trigonometry. Seek out a tutor from your college, honorary societies, or university to help you learn more about that topic.

Follow several of the techniques in this chapter to enhance your learning experience with the tutor. Record your efforts and the results. We are confident that somewhere during this effort you will begin to feel you understand the subject area better than the tutor.

1.6 Conclusion

In this chapter, you have learned some simple techniques that, when implemented, will significantly enhance your classroom experience and in turn lead ultimately to higher grades. An important part of what you have learned is to document your efforts to acquire the skills and the responses, yours as well as others, to those efforts. This documentation will allow you to assess your progress and, like everything in this textbook, allow you to exercise and refine your skills to help you to be a better student and later on enhance your engineering skills.

Also, you need to remember to keep exercising those skills in order to improve them. Once you have fully developed the skill, you will be able to reduce the documentation process, although you should never stop the process entirely. Remember, the star football player is always reviewing game films!

Let us now review the significant points:

1. Asking questions in class is the single most important activity to developing your engineering skills. The most powerful part of this process is your becoming actively involved in the educational process.
2. Join (or form) a study team of at least three members (including yourself). Again, you must be active in this group by asking questions as well as answering them in order to receive maximum benefit.
3. Become active in your student professional organization. Like item 2, the operative word is *actively* involved. Within that activity, you need to both ask questions about your studies and answer questions that may be asked of you.
4. Finally, you need to take advantage of outside-the-classroom resources such as the professor's office hours. Seek out the professor during posted office hours. Seek out other faculty members where possible and appropriate. Look to your honorary societies and university supplied tutors.

Good luck with these activities, they really will help you to work smarter!

END OF CHAPTER REVIEW QUIZ

Select the most appropriate answer to the following statements.

1. The most important way to earn good grades is to
   - **a.** Work a lot of homework problems.
   - **b.** Spend more time reading the course text.
   - **c.** Work smarter.
   - **d.** Make use of the library.
2. The most important activity to help you earn higher grades is to
   - **a.** Ask questions in class.
   - **b.** Study together in study groups or study teams.
   - **c.** Join and be active in your student organization.
   - **d.** Take advantage of outside-the-classroom resources.
3. The most important value of asking questions in class is to
   - **a.** Learn how to formulate a question.
   - **b.** Become actively involved in the educational process.
   - **c.** Ask questions such that you are able to have the professor explain things in a different manner.
   - **d.** Help others learn better.

4. The second most important activity you can do to earn higher grades is to
   a. Ask questions in class.
   b. Study together in study groups or study teams.
   c. Join and be active in your student organization.
   d. Take advantage of outside-the-classroom resources.

5. The advantage of studying together in study groups or study teams is that
   a. Engineers almost always work together in a team.
   b. You learn a lot from hearing other students respond to questions.
   c. You tend to learn more from peers.
   d. All of the above.

6. The third most important activity you can do to earn higher grades is to
   a. Ask questions in class.
   b. Study together in study groups or study teams.
   c. Join and be active in your student organization.
   d. Take advantage of outside-the-classroom resources.

7. What is the most important reason to join and be active in a student organization?
   a. It looks great on your resume.
   b. It helps in learning how to network for both technical and career success.
   c. It helps you to develop leadership skills.
   d. You will significantly enhance your soft skills like public speaking and team building.

8. The fourth most important activity you can do to earn higher grades is to
   a. Ask questions in class.
   b. Study together in study groups or study teams.
   c. Join and be active in your student organization.
   d. Take advantage of outside-the-classroom resources.

9. The most important outside-the-classroom resource of which to take advantage is
   a. Tutors
   b. The classroom instructor during his or her office hours.
   c. Extensive use of the library.
   d. Effective use of the Internet.

10. The most important asset you can have as both a student and a working engineer is
   a. An outstanding skill set.
   b. An outstanding resume.
   c. Outstanding grades.
   d. Lots of activity in your student organization.

EXERCISES TO DEVELOP AND ENHANCE YOUR SKILL SET

Section 1.2 Asking Questions in Class

Exercise 1.1 Develop questions that you can ask in each of your classes. Ask those questions and determine which of your instructors really welcome questions in their classes. If you turn this in as an assignment, make sure you do not report specifics of a particular course.

Exercise 1.2 In those classes from Exercise 1.1, ask questions over the next two weeks and record any improvement you feel you have achieved due to asking questions.

Exercise 1.3 From the classes from Exercise 1.1, which instructor seems to respond best to questions? In your own opinion, why do you feel the instructor does respond so well to questions?

Exercise 1.4 From the classes from Exercise 1.1, which instructor seems to respond most poorly to questions? Why do you feel the instructor does respond so poorly?

Section 1.3 Forming Study Groups or Study Teams

Exercise 1.5 Identify which class is your best choice for developing a study team. Why did you make the choice that you did? Develop a study team for that class.

Exercise 1.6 Do the same as you did in Exercise 1.5 for all the rest of your classes.

Exercise 1.7 Monitor and record your activities in your study team(s) for the next three weeks. What can you deduce from the activities within your study team(s)? Continue this throughout the current academic term.

Section 1.4 Joining and Being Active in Your Professional Organizations

Exercise 1.8 First, realize that you really can choose any student organization in which to learn skills, so if you are not ready to fully determine your major, pick the organizations you would be most interested in. Develop a list of pros and cons for each of these organizations. Then pick the one you will join.

Exercise 1.9 Attend a meeting of your student organization and take notes. What did you learn?

Exercise 1.10 Determine a list of committees or activities available in your organization. Pick the one you like best and join it.

Exercise 1.11 Determine what you will need to do in order to be active in that group and then do it.

Exercise 1.12 Record the results of your being active in that group. Identify the upperclass members of the group who have taken courses in which you are currently engaged. Ask one or more of them to help you form a concept from one of those classes.

Section 1.5 Taking Advantage of Outside-the-Classroom Resources

Exercise 1.13 Identify as many of the outside-the-classroom resources as you can. Then determine the process you need to follow in order to take advantage of them. For example, determine and record the office hours of your current instructors.

Exercise 1.14 Does your department, college, university, or student honorary organizations offer tutoring? Are there charges for these services? If so, how much do they cost?

Exercise 1.15 Take advantage of as many of the services identified in Exercise 1.13, and evaluate their effect on what you are learning in your classes.

CHAPTER

Develop Your Skill Set

> *"Scientists investigate that which already is; Engineers create that which has never been."*
>
> —Albert Einstein

> *"Scientists dream about doing great things. Engineers do them."*
>
> —James A. Michener

2.1 Introduction

In this chapter, we focus on the skills the organization, ABET (the organization that accredits engineering programs in the United States), feels all engineers must have. The importance of these skills cannot be overstated and the student who develops a plan to enhance these will see positive benefits in enhancing his or her educational experience. In addition, your success as an engineer will also be enhanced by how well you develop and enhance those skills during and after the academic part of your career.

ABET is the organization that is responsible for establishing criteria that engineering schools must meet in order for their programs to be academically accredited. It should be noted that ABET accredits only engineering, computer science, and technology programs, not departments, colleges, or universities.

The criteria that ABET uses to determine if a program is deserving of accreditation are skill based rather than knowledge based with an understanding that the acquisition of these skills will lead to a significantly enhanced career as an engineer. In addition, these criteria will also cause engineering programs to continually strive to improve!

> ABET's criteria used in determining whether a program can be accredited is designed to make sure students acquire the skills which will lead to a significantly enhanced career and to make sure engineering programs continually strive to improve!

LEARNING OBJECTIVES

By using the information and exercises in this chapter you will be able to:

- Enhance your ability to speak effectively.
- Enhance your ability to write effectively.
- Enhance your ability to read effectively.
- Enhance your ability to listen effectively.
- Learn how to add structure to your problem solving skill.
- Learn how to identify weaknesses in basic principles of math and science.
- Master the ability to understand and perform experimentation.
- Enhance some of the basic elements that comprise design.
- Increase facility to function effectively on multidisciplinary teams.
- Enhance your ability to identify, formulate, and solve engineering problems.
- Become more aware of the need to understand professional and ethical responsibility.
- Better understand the impact of engineering solutions in a global, economic, environmental, and societal context.
- Understand the importance of lifelong learning.
- Understand the importance of knowing contemporary issues.
- Master the techniques of using all the tools associated with modern engineering practice.

We look at how you can work on developing and enhancing these skills on your own. We use techniques developed as part of the ProSkills© program so that, through repetition, we strive for continuous improvement in as many skills as possible. You will realize that while you work on developing and enhancing these skills your grades will also improve.

The essence of these criteria is captured in what ABET calls *"Criterion.3 Program Outcomes and Assessment."* In this criterion, engineering programs must demonstrate that their graduates have

a. an ability to apply knowledge of mathematics, science, and engineering.

b. an ability to design and conduct experiments, as well as to analyze and interpret data.

c. an ability to design a system, component, or process to meet desired needs within realistic constraints such as economic, environmental, social, political, ethical, health and safety, manufacturability, and sustainability.

d. an ability to function on multidisciplinary teams.

e. an ability to identify, formulate, and solve engineering problems.

f. an understanding of professional and ethical responsibility.

g. an ability to communicate effectively.

h. the broad education necessary to understand the impact of engineering solutions in a global, economic, environmental, and societal context.

i. a recognition of the need for, and an ability to engage in lifelong learning.

j. a knowledge of contemporary issues.

k. an ability to use the techniques, skills, and modern engineering tools necessary for engineering practice.

Mastery of these skills is so important that we feel that they represent a *complete* set! These are the skills that define the successful engineer!

The schools are primarily responsible for making sure you acquire these skills, however, our premise is that you need to make yourself responsible for fully developing and enhancing these skills before you graduate. *Do not rely on your school to develop them for you, only you can do that!* Remember, skill development is like learning mastery over a sport, you cannot read about it and be successful, and you must practice, practice, and practice.

You will take a look at each of these and be given guidance relative to techniques you can use to assure you maximize your skill level in each area. An interesting observation: if you as well as your school are actively doing this, then you clearly will end up with a much better skill set than those students who let the school be totally responsible for the development of their skill set. For all practical purposes, this latter approached is destined to fail. It is like a coach being totally responsible for training a football player who passively stands there without participating in the coach's training program.

Although all of these are important, we will start with item (g) "An ability to communicate effectively" since we strongly feel this is the most important skill an engineer can posses. In fact, this set is so important that it is addressed in other chapters of this book. As Dr. Alexander has frequently stated over the last thirty years: *"Your success as an engineer will be directly proportional to your ability to communicate!"* Now is the time for some important clarification. When we refer to communication skills we are referring to the ability to read, write, and listen as well as the ability to speak. So remember whenever we say communicate we are referring to all four elements!

"Your success as an engineer will be directly proportional to your ability to communicate!"

Mastering communication skills means mastering the ability to

- read,
- write,
- listen, and
- speak effectively.

2.2 ABET Criterion 3(g) "An Ability to Communicate Effectively"

People in industry have repeatedly complained that engineering graduates are ill prepared in written and oral communication. The engineering student who communicates effectively is a valuable asset. You may speak or write easily and quickly. However, how *effectively* do you communicate? The art of effective communication is of the utmost importance to your success as an engineer.

For engineers in industry, effective communication is extremely important for making yourself very valuable to your employer as well as making you promotable. Consider the result of a survey of U.S. corporations that asked what factors influence managerial promotion. The survey includes a listing of twenty-two personal qualities and their importance in advancement. You may be surprised to note "technical skill based on experience" placed fourth from the bottom.

Attributes such as self-confidence, ambition, flexibility, maturity, ability to make sound decisions, getting things done with and through people, and capacity for hard work all ranked higher. At the top of the list was *"ability to communicate."* The higher your professional career progresses, the more you must be able to communicate effectively. Therefore, you should regard effective communication as the most important tool in your engineering tool chest.

Learning to communicate effectively is a lifelong task you should always work to enhance. The best time to begin is right now! Continually look for opportunities to develop and strengthen your reading, writing, listening, and speaking skills. You can do this through classroom presentations, team projects, active participation in student organizations, and enrollment in communication courses. Finally, you can join Toastmasters (they even exist on campuses). Go to http://www.toastmasters.org/.

Although two chapters are devoted to communication skills, we cover them here but not in as much detail.

First, however, we digress for a moment to another topic that is related to what we want you to accomplish. Throughout our career we have focused on helping students do better. One of the most commonly asked question is "What can I do to earn better grades?" Before we go further, why not take a break and write down what you think a student should do to earn higher grades.

The four things you can do to earn higher grades were covered in our chapter on earning higher grades. But, key to all four items, is the use of all the communication skills.

2.2.1 Speak Effectively

This is the most commonly addressed communication skill and yet, rarely does one really develop it appropriately. The most important aspect is to remember that the goal is to communicate what you want to communicate. We emphasize the following repeatedly because it is so important: "The most effective way to learn how to communicate clearly is to first focus on eliminating barriers!" The objective of all communication is to have your message be understood. So ask yourself: Is the way I say something creating a barrier to my message being understood? Is the way I am dressed causing a barrier? Are my slides creating a barrier?

The most effective way to learn how to communicate effectively is to first focus on eliminating barriers.

Let us look at some simple but very powerful rules. Use simple sentences. Keep the structure of the sentences simple. Use easy-to-understand words (even if it takes more words to say or write something, it is much better to use words that people understand without thinking about them). If someone stops listening to admire a word you have used or worse, stops listening because you use a word or acronym they do not understand, then you have failed!

EXAMPLE 2.2.1

Use the "ask questions" part of earning higher grades to practice your speaking skills in a class.

■ Solution

One of the most effective techniques for learning speaking skills in the classroom is to just ask the professor "Is this, what you have stated?" Then respond, in your own words, by repeating what you think the professor said.

PRACTICE PROBLEM 2.2.1

Pick a class and when appropriate, ask the professor "Is this, what you have stated?" At the end of that question, insert your words reflecting what you think the professor said.

Then, at your earliest convenience, write down the professor's response. In addition, you may have actually gotten into an exchange of ideas before you had the response you wanted. Record this as well. Notice, we are exercising your writing skills and listening skills as well.

EXAMPLE 2.2.2

Use your study partners as a proactive vehicle for enhancing your speaking skills.

■ Solution

Summarize the most recent lecture you experienced for your partner in five minutes. Then ask the partner to critique your performance using the form to be used to evaluate speaking skills which can be found on our website.

PRACTICE PROBLEM 2.2.2

Write a summary of one of your previous lectures. Present it to your study partners and obtain feedback using the form from the website.

Another way of enhancing your speaking skills is through active involvement with your professional societies. This is extremely good advice anytime during your career. Clearly participating in committees in these organizations will give you plenty of opportunities for enhancing your speaking skills. The most important thing to keep in mind is that volunteer organizations offer a very low or no-risk practice environment. However, the work environment tends to be a high-risk place to practice your skill set.

EXAMPLE 2.2.3

Find an easy way to practice your speaking skills in a student organization.

■ **Solution**

You choose to join the Society of Automotive Engineers SAE and volunteer to work on the Mini-Baja vehicle redesign committee. You then proceed to find out all you can about previous designs, about what worked and what did not work. You can take the result of this historical information and organize a PowerPoint presentation for the next meeting of the committee. Finally, you present the results of this study and have the committee use the form from our website to evaluate your performance.

PRACTICE PROBLEM 2.2.3

Ask to be assigned to a committee of your student professional organization and then look for something you can do to help the committee by doing some data gathering where you can present the results at a future committee meeting. Prepare a PowerPoint presentation for your committee.

2.2.2 Write Effectively

Annotating your homework and notes may be the single most effective way of enhancing your writing skills. As far as what writing style to use, consider the following:

> To understand the best writing style, subscribe to and carefully read magazines related to your professional organizations and popular magazines such as *Reader's Digest*.

A major problem with writing is that it is almost always viewed as an afterthought and not a partner in our creativity. We now urge you to rethink how you write. Start today to see writing as a method for "thinking through a problem or task." Begin by annotating a homework problem on which you are currently working.

EXAMPLE 2.2.4

Express a problem from one of your courses in words, using no equations, figures, or other graphical forms.

Find a solution for a physics problem involving a spring and a mass and a dampening system. The mass is 1 kilogram, the spring constant is 1 newton per meter, and the damping is 1 newton second per meter. Find the response (meters per second) when there is an initial displacement of 10 centimeters.

PRACTICE PROBLEM 2.2.4

Express a problem from one of your courses in words, using no equations, figures, or other graphical forms.

2.2.3 Read Effectively

Dr. Alexander had a very unique student who could tell much more about a product by what the product specification sheet did not say about the product. Most of us read something trying to understand what is being said in the words before us. In reality, whatever we read needs to be analyzed by what is in the entire document or text. Rarely can the truth be ascertained from a single statement. In fact, we go even further and tell students to use at least two textbooks when trying to learn material. As teachers, many of us try to present our understanding of a topic in our own words and in a different manner from what is presented in the course textbook, forcing the student to try and relate what has been said with what the student has read. If the student cannot relate the two approaches, whether reading two books or listening to a lecture and reading one book, then the student must ask for clarification.

EXAMPLE 2.2.5

Find a paragraph to read and then read it. Next put what you have read into your own words. Finally, validate it from other sections in the text or do so using an external source.

■ **Solution**

As part of a course in circuit theory, a student just read a paragraph out of a circuits textbook that presents Kirchhoff's current law (KCL). The essence of that paragraph states, "The algebraic sum of all the currents flowing out of a closed surface is equal to zero." Note the student stated this in words rather than using an equation, which the student could also have done.

Next, the student looks at another circuits textbook and finds that, "The sum of the current entering a node is equal to the current leaving it." Is this the same statement? No. However, they must both be correct. How do you reconcile the two statements? In the first statement, the student refers to a closed surface. In the second statement, the reference is a node. Obviously you could enclose the node inside a sphere, which is a closed surface. In the first statement, current following out of a closed surface is positive. Current flowing into the closed surface is negative. Taking the negative terms over to the right-hand side of the equation, they now become positive and you have, from the first statement, the total current flowing out of a closed surface is equal to the total current flowing into the closed surface. This is now in agreement with the modified second statement, and you now have a much better understanding of KCL.

PRACTICE PROBLEM 2.2.5

Find a suitable paragraph in one of your textbooks and read it. Put it into your own words. Next, find another textbook with a similar topic. Read the relevant section and compare it with what you read the first time. Do they agree? What do you have to do to make them the same or agree?

2.2.4 Listen Effectively

Effective listening is very important for success in any field, but have you ever taken a course in effective listening? If you have, please let us know, we sure would like to know about it.

The best way to learn how to be a good listener is to focus on listening and then practicing it over and over. Clearly, the process of asking questions in class and then discussing the response with the professor in class involves listening skills enhancement.

One of the best techniques we know of for enhancing listening skills involves just active listening while in a conversation with another individual. Surprisingly, just listening to someone and then engaging that individual in what you are hearing them say will lead the individual to think you are a really great communicator. One of the things we tell individuals is that you have two ears and one mouth for a reason!

The process of listening followed by questions and then interaction leads us to believe that the skill of listening well is relatively easy to enhance.

PRACTICE PROBLEM 2.2.6

Attend a professional meeting and take notes. Share your notes with the individual responsible for the meeting.

This brings up another valuable way to learn how to listen effectively. Volunteer to be the secretary for your professional organization or for any committee in which you participate.

2.3 ABET Criterion 3(a) "An Ability to Apply Knowledge of Mathematics, Science, and Engineering"

As students, you are required to study mathematics, science, and engineering with the purpose of being able to apply that knowledge to the solution of engineering problems. The skill here is the ability to apply the fundamentals of these areas in the solution of a problem. So, how do you develop and enhance this skill? The best approach is to work as many problems as possible in all of your courses. However, if you are really going to be successful with this, you must spend time analyzing where and when and why you have difficulty in easily arriving at successful solutions. You may be surprised to learn that most of your problem-solving problems are with mathematics rather than your understanding of theory. You may also learn that you start working the problem too soon. Taking time to think about the problem and how you should solve it will always save you time and frustration in the end.

What we have found that what works best for us is to apply our six-step problem-solving technique. Then we carefully identify the areas where we have difficulty solving the problem. Many times, our actual deficiencies are in our understanding and ability to use correctly certain mathematical principles. We then return to our fundamental math texts and carefully review the appropriate sections and in some cases work some example problems in that text. This brings us to another important thing you should always do:

Always retain as references, all of your basic mathematics, science, and engineering textbooks.

This process of continually looking up material you thought you had acquired in earlier courses may seem very tedious at first; however, as your skills develop and your knowledge increases, this process will become easier and easier.

EXAMPLE 2.3.1

Pick a problem from one of your engineering courses. Develop a solution using a structured approach. Take the original problem statement and your solution and develop a mapping of the portions that are essentially mathematics and the portions that are engineering science.

■ Solution

The following problem from a basic circuits course has been selected.

Problem statement: For the circuit in Figure 2.1, obtain v_1 and v_2 using nodal analysis (KCL equations).

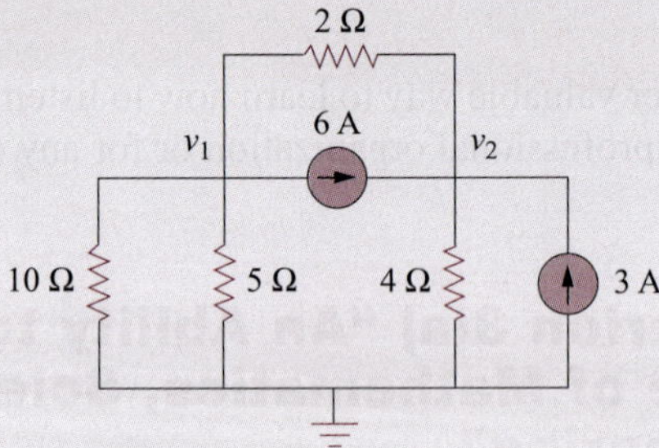

Figure 2.1
Circuit for Example 2.3.1.

Solution to the above problem: The first step for this problem is to write the node equations for the circuit at nodes v_1 and v_2. The next step is to set up the equations so that you can solve for the unknowns. The last step is to actually solve the equations.

At node 1,

You first apply Kirchhoff's current law (KCL) equation at node 1. KCL states, "The algebraic sum of all the currents flowing out of a closed surface must add up to zero." To determine the current flowing out of a closed surface (the node in this case) using Ohm's law, you represent that current as equal to the voltage on the node side of the circuit element minus the voltage on the other side of the element [measured with respect to the reference node (ground)] divided by the resistance of the circuit element. Thus, you get,

$$\frac{v_1 - 0}{10} + \frac{v_1 - 0}{5} + 6 + \frac{v_1 - v_2}{2} = 0 \qquad \textbf{(2.3.1.1)}$$

At node 2,

$$\frac{v_2 - v_1}{2} + (-6) + \frac{v_2 - 0}{4} + (-3) = 0 \quad \textbf{(2.3.1.2)}$$

Note, the currents delivered by the two independent current sources are listed in this equation as negative since they are flowing into the node not out of it.

The next step is to set up the equations that will lead you to a solution. First you will multiply Equation 2.3.1.1 by 10, take the known value over to the right side of the equations, and then combine terms.

$$v_1 + 2v_1 - 60 + 5v_1 - 5v_2 = 0 \text{ or } 8v_1 - 5v_2 = 60 \quad \textbf{(2.3.1.3)}$$

The next step is to set up the equations that will lead you to a solution. First you will multiply Equation 2.3.1.2 by 4, take the known over to the right side of the equations, and then combine terms.

$$2v_2 - 2v_1 - 24 + v_2 - 12 = \text{or } -2v_1 + 3v_2 = 36 \quad \textbf{(2.3.1.4)}$$

You now have two choices, you can use matrix inversion to solve for v_1 and v_2, either performing the calculations by hand or by using MATLAB, or you can just eliminate one of the unknowns by adding Equation 2.3.1.3 to Equation 2.3.1.4 after multiplying Equation 2.3.1.4 by 4.

$$[8v_1 - 5v_2 = 60] + 4[-2v_1 + 3v_2 = 36] \text{ or}$$
$$(8 - 8)v_1 + (-5 + 12)v_2 = 60 + 144, \text{ which leads to}$$
$$7v_2 = 204 \text{ or } v_2 = 29.14 \text{ volts.}$$

You can now insert this value into Equation 2.3.1.3 to obtain v_1.

$$8v_1 - 5(29.14) = 60 \text{ or } 8v_1 = 145.7 + 60 = 205.7 \text{ or}$$
$$v_1 = 25.71 \text{ volts.}$$

The final step of this solution is to identify the parts of this problem that are engineering science and the parts that are basic mathematics. Clearly, the engineering science is where you apply Kirchhoff's current law and Ohm's law. The rest of the solution process is mathematics.

PRACTICE PROBLEM 2.3.1

Pick a problem from one of your engineering courses. Develop a solution, and then take the original problem statement and your solution and develop a mapping of the portions of the problem that are essentially mathematics and the portion that is essentially engineering science.

2.4 ABET Criterion 3(b) "An Ability to Design and Conduct Experiments, as Well as to Analyze and Interpret Data"

Engineers must be able to design and conduct experiments, as well as analyze and interpret data. Most students have spent many hours performing experiments in high school and in college. During this time, you have been asked to analyze the data and to interpret the data. Therefore, you should already be skilled in these two activities. My recommendation is that, in the process of performing experiments in the future, you spend more time in analyzing and interpreting the data in the context of the experiment.

What does this mean? If you are looking at a plot of voltage versus resistance or current versus resistance or power versus resistance, what do you actually see? Does the curve make sense? Does it agree with what the theory tells you? Does it differ from expectation, and, if so, why?

Clearly, practice with analyzing and interpreting data will enhance this skill. Since most, if not all, the experiments you are required to do as a student involve little or no practice in designing the experiment, how can you develop and enhance this skill?

Actually, developing this skill under this constraint is not as difficult as it seems. What you need to do is to take the experiment and analyze it. Just break it down into its simplest parts, reconstruct it trying to understand why each element is there, and finally, determine what the author of the experiment is trying to teach you. Even though it may not always seem so, every experiment you do was designed by someone who was sincerely motivated to teach you something.

PRACTICE PROBLEM 2.4.1

Pick an experiment from one of your engineering, physics, biology, or chemistry courses and decompose it into its basic elements. Then reconstruct the experiment trying to determine what the creator of the experiment might have wanted you to learn. Carefully document what you are doing so that you have a complete record for later.

2.5 ABET Criterion 3(c) "An Ability to Design a System, Component, or Process to Meet Desired Needs within Realistic Constraints Such as Economic, Environmental, Social, Political, Ethical, Health and Safety, Manufacturability, and Sustainability"

The "ability to design a system, component, or process to meet desired needs" is why engineers are hired. That is why this is the most important *technical* skill that an engineer has. Interestingly, your success as an engineer is directly proportional to your ability to communicate, but your being able to design is why you will be hired in the first place. Design takes place when you have what is termed an open-ended problem that eventually is defined by the solution.

Within the context of this course or textbook, we can only explore some of the elements of design. Pursuing all of the steps of our problem-solving technique teaches you several of the most important elements of the design process.

Probably the most important part of design is clearly defining what the system, component, process, or in our case, problem is. Rarely is an engineer given a perfectly clear assignment. Therefore, as a student, you can develop and enhance this skill by asking yourself, your colleagues, or your professors questions designed to clarify the problem statement.

Exploring alternative solutions is another important part of the design process. Again, as a student, you can practice this part of the design process on almost every problem you work.

Evaluating your solutions is critical to any engineering assignment. Again, this is a skill that you, as a student can practice on every problem you work.

Design is really the creative part of being an engineer. Many feel that design is natural talent and if you have it, you can become a successful engineer, if you do not have it you can never become a successful engineer. In addition, those individuals are, most likely, the same individuals that believe you cannot "teach" design!

We feel that both of these viewpoints are false. Let us address the first. Our viewpoint is that even if you are not born with the ability to design, if you are intelligent and motivated to become an engineer, you can acquire the ability to design!

> Design is a skill like anything else and you acquire competence in design the same way you acquire any other skill. Learning what to practice and then practicing, practicing, and practicing is what it is all about!

Think of design the way you would paint a picture. Someone asks you to design something. That is like asking you to paint a picture of a barn or a landscape. Once you have the higher level description of what you want to do, you begin breaking up the problem (painting) into manageable subcomponents. You build the subcomponents into a working module and then combine them into an integrated whole.

We have found that the best way to build design skills in undergraduates is to have them practice simple design activities—the sum of which various components make up the design process. One of the best is to have them design a simple experiment in an undergraduate laboratory.

EXAMPLE 2.5.1

Design a spring, mass, and damping system that demonstrates underdamped, critically damped, and overdamped responses to step inputs.

■ **Solution**

This can be easily done with a number of software packages as well creating your own software solution. And, that is a perfect acceptable solution for this type of problem. However, it might be more fun and helpful if you go to your physics department or mechanical engineering department and see if they have a simple physical system that will allow you to control the friction of a simple spring and mass system. You do not need to be overly creative. You can start by setting the friction to zero and simply moving the mass a reasonable distance and letting go and record what happens.

Next you could set the value of the friction that results in five or six oscillations before the mass comes to rest. The oscillations will not be observable until you displace the mass as you did in the zero friction case.

Another thing to observe is what value of friction causes the mass to return to near its rest position after displacement, with no oscillations. Once the mass stops moving in this case, is it at the original rest position? If it is not, why did it not return to exactly the same position?

PRACTICE PROBLEM 2.5.1

Design an experiment that relates an *R*, *L*, and *C* circuit to a spring, mass, and damping system.

Another neat way to enhance this skill is to design problems for an engineering class in which you are currently enrolled. For example, if you are in a statics class, you might design a problem that uses the conservation of forces to find an unknown force.

PRACTICE PROBLEM 2.5.2

Design a problem or experiment for an engineering class in which you are currently enrolled.

2.6 ABET Criterion 3(d) "An Ability to Function on Multidisciplinary Teams"

The "ability to function on multidisciplinary teams" is inherently critical for the working engineer. Engineers rarely, if ever, work by themselves. Engineers will almost always be part of some team. One of the things we like to remind students is that you do not have to like everyone on a team; you just have to be a successful part of that team.

Most frequently, these teams include individuals from of a variety of engineering disciplines, as well as individuals from nonengineering disciplines such as marketing and finance. Students can easily develop and enhance this skill by working in study groups in every course they take. Clearly, working in study groups in nonengineering courses as well as engineering courses outside your discipline will also give you experience with multidisciplinary teams.

The "ability to function on multidisciplinary teams" is inherently critical for the working engineer. Engineers rarely, if ever, work by themselves. Engineers will almost always be part of some team.

PRACTICE PROBLEM 2.6.1

Pick two classes in which you are currently enrolled. Form a study group for each class that is distinct from each other. Document your activities, successes, and failures over a period of four weeks.

Another way to enhance this skill is to volunteer to serve on a committee in one of your student organizations. The real advantage with this approach is that it is one of limited risk. This is just another one of the many important reasons why you should join and be active in your student organizations.

PRACTICE PROBLEM 2.6.2

Go to the chair or president of your student organization and volunteer to serve on a committee or task force. Document your activities for a period of two months.

2.7 ABET Criterion 3(e) "An Ability to Identify, Formulate, and Solve Engineering Problems"

Much of what you have already done in this chapter addresses many elements of this criterion. In many ways, it is this approach that helps you to learn how to design, and so on. What will help in this is to follow the systems approach as developed in our chapter on using the systems approach to engineering. The process is readily evolved into a general approach for engineering problems. You may be pleased to learn that this process works well for nonengineering courses and, in general, for many important nonengineering activities as well.

PRACTICE PROBLEM 2.7.1

Briefly read our chapter on using the systems approach to engineering and develop a relationship between the systems approach and how each activity relates to *identifying engineering problems, formulating engineering problems,* and *solving engineering problems.*

2.8 ABET Criterion 3(f) "An Understanding of Professional and Ethical Responsibility"

"An understanding of professional and ethical responsibility" is required of every engineer. To some extent, this understanding is very personal for each of us. Let us identify some pointers to help you develop this understanding.

One of our favorite examples is that an engineer has the responsibility to answer what we call the "unasked question." For instance, assume that you own a car that has a problem with the transmission. In the process of selling that car, the prospective buyer asks you if there is a problem in the right-front wheel bearing. You answer no.

However, as an engineer, you are required to inform the buyer that there is a problem with the transmission without being asked.

Your responsibility both professionally and ethically is to perform in a manner that does not harm those around you and to whom you are responsible. Clearly, developing this capability will take time and maturity on your part. You can practice this by looking for professional and ethical components in your day-to-day activities. We refer you to our chapter on ethical dilemmas for more details regarding this important skill development.

Engineers have the responsibility of answering the unasked question!

Remember when we referred to the importance of being a valuable member of a team? No engineer can thoroughly check everyone's work. So there needs to be an element of trust that you have in your team members so you can comfortably and confidentially accept and use their results. This is where ethics comes in big time!

If a member of your team has questionable ethics, then how can you trust their work?

2.9 ABET Criterion 3(h) "The Broad Education Necessary to Understand the Impact of Engineering Solutions in a Global, Economic, Environmental, and Societal Context"

As a student, you must make sure you acquire "the broad education necessary to understand the impact of engineering solutions in a global and societal context." To some extent, if you are already enrolled in an ABET-accredited engineering program, then some of the courses you are required to take must meet this criterion. Our recommendation is that even if you are in such a program, you look at all the elective courses you take to make sure that you expand your awareness of global issues and societal concerns. The engineers of the future must fully understand that they and their activities affect all of us in one way or another.

PRACTICE PROBLEM 2.9.1

Go through the list of courses that is offered in your university. Try to find four courses that, if you were to take them, would help you to better understand the impact of engineering solutions in a global and societal context.

Document the courses and the reasons why you listed them. Later, if you do take these courses, document what you have learned relative to ABET Criterion 3(h).

2.10 ABET Criterion 3(i) "A Recognition of the Need for, and an Ability to Engage in Lifelong Learning"

You must be fully aware of and recognize the "need for, and an ability to engage in life-long learning." It almost seems absurd that this need and ability must be stated. Yet, you would be surprised at how many engineers do not really understand this concept. The only way to be really able to keep up with the explosion in technology you are facing now and will be facing in the future is through constant learning. This learning must include non-technical issues as well as the latest technology in your field.

PRACTICE PROBLEM 2.10.1

An unfortunate reality with many of the classes you take is that you end up hating learning and, in some cases, vow not to "take another class, open another book, nor take another exam!"

Identify the course that you best associate with these negative feelings. Within that course, identify those things that you feel would be interesting to learn more. Pursue learning as much about those things as you can even though learning more about them may *not* help your grade in that course. Document your success and what you learned about yourself and how you learn! Can you relate this to the career plan you developed in our chapter on career planning?

The best way to keep up with the state of the art in your field is through your colleagues and association with individuals you meet through your technical organization or organizations. Reading state-of-the-art technical articles is the next best way to stay current.

You are urged to study our chapter on networking and professional development skills to develop these skills more fully.

2.11 ABET Criterion 3(j) "A Knowledge of Contemporary Issues"

One of the stereotype images of an engineer, is that engineers know nothing about anything other than engineering. This image is totally inaccurate. Successful engineers tend to be well informed about the world around them. The biggest reason for this is that successful engineers have to be since almost everything they do is part of the real world.

This is going to be even truer in the twenty-first century. To have a truly meaningful career in this century, you must have knowledge of contemporary issues, especially those that may directly affect your job and/or work. One of the easiest ways to achieve this is to read a lot; newspapers, magazines, and contemporary books. As a student enrolled in an ABET-accredited program, some of the courses you take will be directed toward meeting this criterion.

PRACTICE PROBLEM 2.11.1

Form a current events group where each member of the group is to bring at least two contemporary issues to the group meeting. The group should plan to meet once per week for 2 hours.

Do this for two months. Carefully document everything you learn. Once you have completed this, can you now develop a plan to continue this activity?

2.12 ABET Criterion 3(k) "An Ability to use the Techniques, Skills, and Modern Engineering Tools Necessary for Engineering Practice"

The successful engineer must have the "ability to use the techniques, skills, and modern engineering tools necessary for engineering practice." Clearly, a major focus of this textbook is to do just that. Learning to use skillfully the tools that facilitate your working in a modern "knowledge-capturing integrated design environment" (KCIDE) is fundamental to your performance as an engineer. The ability to work in a modern KCIDE environment requires a thorough understanding of the tools associated with that environment.

PRACTICE PROBLEM 2.12.1

Review the criteria a through k. Make a list of additional skills an engineer might need.

The successful engineer, therefore, must keep abreast of the new design, analysis, and simulation tools. That engineer must also use those tools until he or she is comfortable with using them. The engineer also must make sure software results are consistent with real-world actualities. It is probably in this area that most engineers have the greatest difficulty.

Successful use of the *techniques, skills, and modern engineering tools necessary for engineering practice* requires constantly learning and relearning the fundamentals of the area in which the engineer is working.

2.13 Conclusion

In this chapter, you have learned a great deal about the skills you need to master in order to have a successful career. You have learned some valuable techniques to enhance your communication skills.

You have learned how important it is to have well-developed experimental skills. You have learned how to develop and enhance these skills through designing experiments and performing experiments.

You have been exposed to the importance of good, structured problem-solving techniques and what those techniques are. Through this process you have become aware of how to identify weaknesses in the basic principles of math and science.

In the area of design, you have identified elements that you can practice in order to enhance your design process skills. These, along with the other items you have identified, will help you become a successful design engineer!

An increased facility to work in multidisciplinary teams is now a part of your skill set. This, along with a great awareness of the importance of knowing contemporary issues, will make you a more valuable employee.

You are now fully aware of the importance of mastering all the tools of your profession. These are all the tools associated with modern engineering practice.

Finally you are now aware of the importance of lifelong learning and how important that is to your success. Along with the importance of conducting yourself in a professional and ethical manner, this can bring it all together for you.

END OF CHAPTER REVIEW QUIZ

Select the most appropriate answer to the following statements.

1. ABET is the agency that
  a. Determines the accreditation of university,
  b. Determines the accreditation of colleges of engineering,
  c. Determines the accreditation of engineering, computer science, and technology programs,
  d. Determines the accreditation of engineering, computer science, and technology departments,

2. ABET criteria is used to determine if a program is
  a. Deserving of accreditation.
  b. Continually improving.
  c. Making sure their students have the necessary skills to be successful engineers.
  d. All of the above.

3. The ABET criteria is based
  a. On skills.
  b. On a minimum number of mathematic courses.
  c. On a minimum number of social science courses.
  d. All of the above.

4. Although skill development is the primary focus of ABET criteria, what knowledge is expected
  a. Mathematics.
  b. Science.
  c. Engineering.
  d. All of the above.

5. What do we consider to be the most important soft skill an engineer needs to develop and enhance?
  a. Working in teams
  b. Communication (reading, writing, listening, and speaking)
  c. An understanding of society
  d. An understanding of politics

6. The most important engineering skill is
   a. The ability to do design.
   b. The ability to analyze engineering systems.
   c. The ability to do research.
   d. The ability to start your own business.
7. As a working engineering you are responsible to understand the impact of what you are doing on
   a. Economics.
   b. Public health and safety.
   c. The environment.
   d. All of the above.
8. What criterion deals with what an engineer needs to engage in throughout their life?
   a. Experimentation
   b. Learning
   c. Active participation in Toastmasters
   d. Networking
9. In what kind of teams should an engineer expect to work ?
   a. Multidisciplinary
   b. A single engineering discipline
   c. Ones composed of only engineers
   d. Ones that are composed of individuals who really like to work with each other
10. As a practicing engineer, you should conduct yourself in a
   a. Confident manner.
   b. Knowledgeable manner.
   c. Professional and ethical manner.
   d. Way in which you make sure you are doing well.

EXERCISES TO DEVELOP AND ENHANCE YOUR SKILL SET

Section 2.2 ABET Criterion 3(g) "An Ability to Communicate Effectively"

Exercise 2.1 Identify three activities in which you will engage during this term designed to develop your ability to read with understanding. Record and analyze what you have learned while engaging in these activities.

Exercise 2.2 Identify three activities in which you will engage during this term designed to develop your ability to write effectively. Record and analyze what you have learned while engaging in these activities.

Exercise 2.3 Identify three activities in which you will engage during this term designed to develop your ability to listen effectively. Record and analyze what you have learned while engaging in these activities.

Exercise 2.4 Identify three activities in which you will engage during this term designed to develop your ability to speak effectively. Record and analyze what you have learned while engaging in these activities.

Section 2.3 ABET Criterion 3(a) "An Ability to Apply Knowledge of Mathematics, Science, and Engineering"

Exercise 2.5 You are and will be doing this in almost all of your engineering, math, and science courses. Pick a course you are taking this term, preferable an engineering course, and identify the math, science, and engineering content of one or more of your assignments.

Exercise 2.6 From one of the courses you are taking, evaluate assignments and/or test where you made errors. Try to identify if the main problem with the errors was in your understanding of mathematics, science, or engineering. Develop a plan to correct the weakness identified. If possible obtain feedback from the course instructor as to the merits of your plan.

Exercise 2.7 Develop a plan to be used in all your current classes and future classes that will ensure your ability to work effectively with mathematics, science, and engineering.

Section 2.4 ABET Criterion 3(b) "An Ability to Design and Conduct Experiments, as Well as to Analyze and Interpret Data"

Exercise 2.8 Develop your own plan for developing and enhancing your "experimental technique." Elements of this plan must contain how you will work with your instructors to develop a clear understanding of what is being asked of you as to the work you will do in the laboratory. Next you should have a plan laying out the process you will follow relative to the experiment including modules that will need to be assembled and analyzed. The process you should follow in putting the experiment you wish to complete together, try assembling it in modules and test each module as you realize it.

Exercise 2.9 Obtain all the elements you will need to assemble to successfully perform your experiment and lay out the equipment you will need. As in Exercise 2.8, build your experiment in modules, testing as you go along, and then assemble the modules into the final experiment and test the completed project.

Exercise 2.10 Collect all required data, making sure you follow procedures that will ensure accuracy in data collection. Carefully review the data and see if they are reasonable. If some of your data do not seem reasonable try to determine why.

Exercise 2.11 Complete your experiment by analyzing the data you collected and determine if the data meet expected results. Explain your results.

Exercise 2.12 Review everything that occurred during the experiment and develop a plan to enhance your experimental technique in the future.

Section 2.5 ABET Criterion 3(c) "An Ability to Design a System, Component, or Process to Meet Desired Needs within Realistic Constraints Such as Economic, Environmental, Social, Political, Ethical, Health and Safety, Manufacturability, and Sustainability"

Exercise 2.13 Although a senior or a graduate student will be doing a lot of this in their senior design and thesis, respectively, elements of this can be practiced by even freshmen students. Go to one of your engineering instructors and work with them to develop some assignments that will enhance your skills.

Section 2.6 ABET Criterion 3(d) "An Ability to Function on Multidisciplinary Teams"

Exercise 2.14 Again, as in Exercise 2.13, seniors and graduate students will have a lot of opportunity to do this. As a student, analyze the study teams you are on for the degree of diversity. What can you do in these teams to increase diversity?

Exercise 2.15 Go to your student organization and evaluate the diversity of the various teams and committees the organization has, and, where necessary, develop suggestions for increasing their diversity.

Exercise 2.16 Gain experience of working on diverse teams or committees. A good source of such groups might be your university-level umbrella student organizations.

Section 2.7 ABET Criterion 3(e) "An Ability to Identify, Formulate, and Solve Engineering Problems"

Exercise 2.17 You should already be doing this in your engineering classes. Pick a class where you are doing this and outline the process you are following. Could your process be improved? What are the identifiable risks in the process you are using? What can you change in your process to mitigate these risks?

Exercise 2.18 Eventually you will learn a lot more about these skills in your senior design projects. Take a survey of the process that will be followed in your senior design projects. Do you see obvious ways this process can be improved? If you do, consider discussing ways to improve the process with one or more of the faculty members involved in the senior design courses.

Section 2.8 ABET Criterion 3(f) "An Understanding of Professional and Ethical Responsibility"

Exercise 2.19 Take a survey of the courses within your academic department that deal with ethics. Analyze the content in the courses that deal with these issues. Do you feel they meet the spirit of this ABET criterion? Do you have any suggestions for improving this content? If you do, consider discussing these with the relevant faculty members.

Exercise 2.20 Develop your own plan for improving your understanding of your professional and ethical responsibility.

Section 2.9 ABET Criterion 3(h) "The Broad Education Necessary to Understand the Impact of Engineering Solutions in a Global, Economic, Environmental, and Societal Context"

Exercise 2.21 When you join an academic department and select the major you will pursue, you will have what is called a program of study, which will contain the courses you will need to successfully complete before you can graduate. If you already have, analyze the courses in your program of study and determine what content will meet this criterion. If you have not yet selected a major, pick one that you might eventually choose and analyze the courses in that program of study and determine what content will meet this criterion.

Exercise 2.22 Analyze what you determined in Exercise 2.20 and see if it is lacking in any way. If you find it lacking, discuss this with an advisor in the department which is responsible for the program.

Exercise 2.23 Since this is an important part of being an engineer, develop a plan to enhance this criterion in outside of the classroom activities. Additionally, extend this plan to include what you will be doing once you graduate.

Section 2.10 ABET Criterion 3(i) "A Recognition of the Need for, and an Ability to Engage in Lifelong Learning"

Exercise 2.24 Take a moment and write down your feelings about learning. Many students do have a "burn out" and develop a very negative attitude to learning. If you have such feelings, develop a plan to change your feelings so that you eliminate a barrier to lifelong learning.

Exercise 2.25 Develop a plan to be followed once you graduate that will ensure that you will continue to engage in lifelong learning and actually to begin to enjoy it.

Section 2.11 ABET Criterion 3(j) "A Knowledge of Contemporary Issues"

Exercise 2.26 Analyze your possible program of study to identify courses that will help you develop your skill in this area. Although courses can help you identify ways to become knowledgeable of contemporary issues, the most effective way to do this is to gain such information from up-to-date print material, the Web, television, your professional organizations, and colleagues. Develop a plan to follow throughout your career to make sure you acquire such knowledge.

Section 2.12 ABET Criterion 3(k) "An Ability to use the Techniques, Skills, and Modern Engineering Tools Necessary for Engineering Practice"

Exercise 2.27 For the program in which you are enrolled or might like to enroll in, identify the techniques, skills, and modern engineering tools you will need to be successful. Does your program of study contain enough course content to acquire all of these skills? If not, where are the weaknesses? Once you have identified these, discuss them with an advisor for that program of study.

Exercise 2.28 As part of your career after graduation, develop a plan to continue to develop and enhance these skills.

CHAPTER

Time Management Skills

Enjoy the Time of Your Life

> *"Time! Every day, you are given 24 hours. To a great extent, you have the power to determine how to invest 1,440 minutes each day. Successful use of 86,400 seconds in 24 hours is directly proportional to the degree of your planning and control."*
>
> —Judith Jurgens, Career Management Consultant

TIME MANAGEMENT SCENARIO

"You are an engaged student who needs to consider not only class schedules, but also extracurricular project deadlines, and work-distribution. You know it is a prudent choice to be involved in multiple facets of the college-engineering environment. Working on technical extracurricular teams and research projects is extremely beneficial, along with nontechnical leadership roles. The benefits of participating in such activities include application of theoretical studies, opportunities to make professional connections, experience in advanced engineering work, learning concepts not taught in the classroom, and building a strong resume.

However, these benefits do not come without cost. You must learn to prioritize and maintain a balance among your activities (social life included). The main challenges you are often presented with are devoting time to tasks that need to be done, as opposed to the tasks you want to do, and where to draw the line between having too many extracurricular activities, and not having enough."

(Courtney Gras, IEEE Student Chapter President, Robotics Team Leader, and Undergraduate Student at the University of Akron)

Do these comments from Courtney Gras sound familiar? After discussing principles of time management, we will visit with Courtney at the end of this chapter to see how she handles this challenge.

3.1 Introduction

Time can be your best friend or worst enemy. You cannot create time, but you can misuse or waste it. So, let us look at how you can invest your time wisely.

LEARNING OBJECTIVES

By using the information and exercises in this chapter you will be able to:

- Understand why structured time use is valuable.
- Evaluate options for time investment.
- Select a successful time management system.
- Implement a time management system that works.
- Slay "dragons" of defeat.
- Make time management easy and effective.
- Maximize career success and enjoy life.

A time management process is more than entering information in a calendar. This is only the first step. Successful time management includes estimating time requirements for activities and developing a schedule to start and complete them by the desired target date.

This chapter will show you how to use proven methods of time management to increase your level of success. Time management plays a leading role in school, in future work situations, and in your personal life. Managing your activities will lead to more accurate work, fewer time pressures, and greater overall results.

If you are already using a time management system, the information in this chapter will verify that you are on target or show you new ways to make your system easier to use and more effective.

If you have tried using time management but have been discouraged with results, this chapter will tell you how to make your future time management efforts more successful.

And, if you have never tried a structured approach to managing time, the concepts in this chapter will give you a step-by-step process to begin a time management system that will work for you.

> *"Successfully managing time is the most beneficial skill that you will need to master early in your career to become successful in today's competitive job environment. Attending meetings, conducting research, participating in group collaboration, and presenting your ideas to others will take up far more time than you can possibly envision right now. Today's graduates are expected to be engineers, marketers, trainers, project managers, presenters, and managers, so bad time management can and will turn a 40 hour work week into an 80 hour work week."*
>
> (Vishnu Pandey, Masters Degree in Electrical Engineering and Systems/Software Engineer)

3.2 Why Time Management is Important to You

As a student, your life is very busy. Very quickly, you will graduate from college and start your working career. You will enjoy greater success if you start now to control how you use time.

To be successful, you need to be in control of your career, and this includes planning and controlling how you use time. While it is impossible to predict detours and other surprise events, you can control how you use time by developing a successful time management system.

In this chapter, we will focus on a systematic approach to planning your time. By using this method you can define what you want to do and get it done on schedule, which will help you achieve higher grades and give you more time for recreation and leisure. It will also reduce stress by showing you how to complete a multitude of activities on time.

3.3 Time Use and Management Resources

Although you invest time in many different ways and for a variety of activities, some activities are more important than others. The time management process starts by first defining how you presently use time.

Start by knowing how you use time.

3.3.1 Time Use

Let us consider typical time needs. A reasonable amount of time is required to be healthy. This includes 7 to 8 hours of sleep each night. Another 2 hours are usually invested in eating and personal care. Recreation contributes to good health, so it is reasonable to allocate some time for leisure activities.

Much of your time is structured. For example, college courses require 48 hours or more each week for class and study. Your professional working career will require at least 40 hours per week on the job. In addition, you will have activities outside the classroom or office that are directly related to your education or work.

In addition to your structured activities, you typically have a balance of discretionary time of 6 to 8 hours each day. The greatest opportunity to control time is associated with those 6 to 8 hours and this chapter will help you manage them.

3.3.2 Time Management Resources

Tools are available to help manage time. Universities typically provide assistance to help you design a time managing system. Start with resources available on campus. Check your university website, or ask your advisor to help you identify the appropriate department.

Books and other resources are available to give you suggestions for successful methods to manage time. Also, many tools are available to help you personalize your time management system. These range from pencil, paper, calendars, and time schedules to more sophisticated tools such as software for your computer and handheld wireless devices.

The Internet also opens a worldwide source of information on this topic. Search on the key words of time management and you will find many uniform resource locators (URLs).

However, the focus of this chapter is not on tools, but on fundamentals that can be applied to any time management system. The secret of successful time management is to apply these principles to whatever system you select. The discussion that follows is based on practical and proven principles to manage time.

3.4 Principles of Managing Time—The "To-Do" List

Successful time management starts by identifying what needs to be done. This is often called a to-do list. If too many activities are included in the list, the first impact may be overwhelming, and you may feel like the person in Figure 3.1.

Time management is less complicated and you can complete activities on time when you use a systematic approach to organizing activities. Don't be overwhelmed by the number of activities on your to-do list. Making your list is just the beginning of the process.

Fortunately, the logical process used by engineers to solve problems is also a useful tool to determine how to handle multiple activities and complete them on time with less stress. Let us consider a different approach to the to-do list by making it a TO and DO process.

Figure 3.1
Typical reaction to overwhelming "To-Do" list.

3.4.1 "TO"—Target and Organize Activities by Due Date and Priority

Start the **TO** process by identifying activities and due dates for the next day, week, and month. Typical activities include class assignments, projects, professional and social activities, recreation, personal errands, and special events. Omit routine activities because they are repetitive and do not require reminders to accomplish.

For major projects, list several smaller steps to be completed. For example, if a written report is due in two weeks, list the outline of the report as the first subactivity. Then repeat the process for the first draft, and continue identifying subactivities that lead to the final product.

Next, select a level of importance for each activity. The following is an effective method to identify levels of importance:

A. Very important and valuable
B. Less important but still valuable
C. Desirable, but not very important

Not all activities have the same value.

Arrange activities by due date and level of importance. For each date, "A" activities should be listed first and then "B" and "C" activities. A sample of the **TO** process is shown in Figure 3.2.

Time Management "TO" Example		
Activities	Due Date	ABC
Send a message to parents	4/5	B
Work homework problems in statics	4/6	A
Meet with career center to discuss resume	4/6	A
Attend baseball game on campus	4/6	C
Read text and prepare for Electronics 1 lecture	4/7	A
Call parents and ask for $50 for car repair	4/7	B
Study for Calculus quiz	4/8	A
Write lab report for electronics class	4/8	A
Attend away soccer game	4/8	C
Send resume to Allied Industries for intern position	4/9	A
Meet with study group for statics homework problems	4/9	A
Work on Baja Race Car with SAE team	4/9	B
Call Allied Industries to check on intern position	4/10	A
Attend IEEE meeting to campaign for Vice President	4/10	B
Change oil in car and check brakes	4/10	C

Figure 3.2
Typical activities with due dates and level of importance.

3.4.2 "DO"—Determine Order of Activities by Time Requirements and Start Dates

To complete the **DO** part of the process, estimate the amount of time needed to complete each activity. Then, define start dates by balancing the amount of time needed with the time available. Results of time allocation and start dates are shown for our example in Figure 3.3.

Often, the start date is one or two days before the due date. In some cases, the start date is the same as the due date if it is associated with a scheduled activity such as a baseball game.

If the total amount of time needed to complete all activities identified for a start date is unreasonable, modify your plans. In the example shown in Figure 3.3, the three activities scheduled to start on April 7 total 9 hours. Options to make this more realistic include preparing for the Calculus quiz or writing the lab report before April 7, or not attending the soccer game.

Tomorrow's "A" activities are more important than today's "C" activities.

Before starting "C" activities, review "A" and "B" activities for the next day. Unless you can complete all activities as planned, it is better to work on some "A" and "B" activities scheduled for the next day before investing time in "C" activities. Until you develop a successful time management system, consider minimizing "C" items. As you gain experience, you will find time to include more "C" activities. This is one of the many benefits of an effective system.

Time Management "DO" Example				
Activities	Due Date	ABC	Hours to Complete	Start Date
Send a message to parents	4/5	B	0.5	4/5
Work homework problems in statics	4/6	A	2.5	4/5
Meet with career center to discuss resume	4/6	A	1.0	4/6
Attend baseball game on campus	4/6	C	2.0	4/6
Read text and prepare for Electronics 1 lecture	4/7	A	2.0	4/6
Call parents and ask for $50 for car repair	4/7	B	0.2	4/6
Study for Calculus quiz	4/8	A	3.0	4/7
Write lab report for electronics class	4/8	A	2.0	4/7
Attend away soccer game	4/8	C	4.0	4/7
Send resume to Allied Industries for intern position	4/9	A	0.5	4/9
Meet with study group for statics homework problems	4/9	A	2.0	4/9
Work on Baja Race Car with SAE team	4/9	B	2.0	4/9
Call Allied Industries to check on intern position	4/10	A	0.5	4/10
Attend IEEE meeting to campaign for Vice President	4/10	B	1.5	4/10
Change oil in car and check brakes	4/10	C	2.0	4/10

Figure 3.3
Activities with estimated time requirements and start dates.

3.4.3 Complete Activities

The best **TO-DO** list will be worthless unless you use it to complete activities. The value of selecting start dates is to ensure activities are completed by their due dates without the need for stressful last minute work sessions.

Start and Finish Activities by Start Date and Priority The most important part of time management is how the system is implemented. Start by working on the activity of highest priority for each day. *Focus on this activity until you complete it.* Do not look at other activities on the list or "take a break" to work on an activity of less importance until you complete the first one.

After finishing an activity, relax a bit. Then go to the next activity and complete it. Don't be discouraged if you do not complete everything exactly as planned. There may be valid reasons why this happens. However, if you do not complete most activities as planned, you may need to revise how you estimate time requirements.

Focusing on one activity at a time will result in more time to accomplish the total list of things to do. You will also enjoy the reward of more time for additional activities by this approach.

Minimize distractions.

If an activity cannot be completed in a reasonable amount of time, you may not have defined it properly or perhaps were distracted and wasted time. Most activities should be completed in an hour or so. Divide large projects into smaller parts. If you have a difficult time focusing on an activity, find a better place to work and minimize distractions.

Document Appropriate Results Record the results of completed activities that relate to your education and career. Some of these can be added to your personal portfolio that is discussed in the chapter on information capturing.

For best results, document activities on a daily or weekly basis. It is easy to forget this part of time management, so start now to keep good records.

Reflect on Results and Enjoy a Reward After you complete a few activities, give yourself a reward. Stop for a moment and reflect on your accomplishments. If you are happy with results, take a break and relax.

When you manage time effectively, you will enjoy many benefits. You will learn more, get higher grades, and minimize stress. With more free time, you can also enjoy doing more of your "C"activities for recreation and fun.

Revise Activity List For maximum results, revise the list of activities on a routine basis. There is no right or wrong time for looking at your plan. Tailor it to fit your lifestyle, but do it every day. If you are a "morning" person, review and revise your list before starting your day. If you are more of an "evening" person, revise your list later in the day. Follow your established procedure to select due dates, to define the level of importance, to estimate the amount of time for completion and to select start dates. This will result in a TO-DO process that will help you complete activities on time.

3.5 Beware of "Dragons"

The "Dragons" identified in Figure 3.4 can destroy the best of time management plans. They can drag you down and defeat your time management efforts. Let's discuss these common "dragons" so you can prevent them from impacting your plan.

Figure 3.4
Time management "dragons."

3.5.1 Procrastination

> *"Procrastination is to defer action, to put off till another day or time, to defer, or to delay."*
>
> (The American College Dictionary)

Procrastination is the number one reason for failing to have a successful time management system. You should never be too busy that you cannot find time to plan. A few minutes invested in the planning process will eventually save you hours of wasted time.

If you are not planning your time, you may be a victim—a victim of the "sorry circle," as shown in Figure 3.5. The only way to break this circle is to plan.

3.5.2 Complicated System

The second dragon is selecting a system that is too complicated or one that does not fit your needs or lifestyle. At first glance, some processes are very appealing. However, if it takes more time to prepare your time management plan than it takes to actually do the activities, you will soon become discouraged and give up.

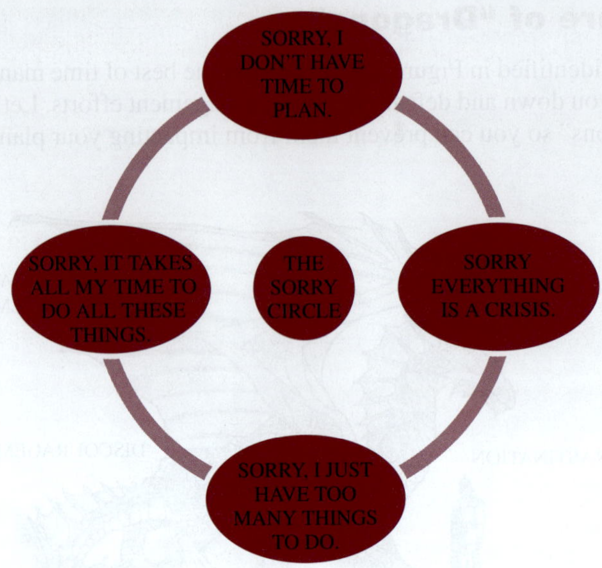

Figure 3.5
The "sorry circle."

3.5.3 Discouragement

And that brings us to the third dragon—becoming discouraged and giving up too soon. If early efforts to control your time investment fall short of your expectations, you may become discouraged.

Implementing a successful process takes time and work. Until you gain experience, it may take several weeks before you are comfortable with your system. So, do not give up too soon, even if the system is not working well in the beginning.

If time management feels like a burden, you have two options—simplify your existing process or consider a different one. Find a system that works for you and then time management will become part of your life. You will accomplish what you want and have more time to do things you just never had the time to do before.

3.6 Making Time Management Effective

Although important to success, time management may not be the most exciting part of your career. It sounds difficult and can become boring. But you have the ability and options to make it much more interesting as well as effective.

3.6.1 Select Resources

Select a time management tool that fits your lifestyle.

Figure 3.6
Time management resource example.

The most successful time management system is one that is simple, easy to use, flexible, and one that works for you. Figure 3.6 shows an example of an electronic resource to track activities.

The resource you select forms the foundation for determining where and how changes can be made to maximize time use. Your time management system should not be a burden, but should be a tool to improve time use and result in more time for pleasure and personal activities.

If time management is new to you, consider talking to other students about how they plan their time. You can do this in your study groups, classes and labs, and by talking to active members of professional organizations. Students who are leaders in organizations typically use time management so they can be involved in many activities and still complete their school work.

The selection and use of any time management process can be fun and successful if you find an approach that fits your style. Experiment with various systems to find one that does not take too much time or effort to use. It is best to keep the process as simple as possible.

Apply time management principles to your specific system.

If you do not presently have a time management system, a good approach is to use the Excel form shown in Figure 3.10. This is an effective way to list activities, identify due dates, assign levels of priority, estimate time requirements for completion, and assign start dates.

The Excel format requires no additional monetary investment, and is a good way for you to become familiar with the process. Then, if you would like to try other time management tools, you can use your experience with Excel to maximize their value.

If you are already using a time management system, you can apply the principles discussed in this chapter to improve results. This is a good time to review the effectiveness of your system. If it is working well, continue using it. If you are not meeting deadlines, use information in this chapter to fine-tune your process.

3.6.2 Use Proven Techniques

If your time management system is effective, you know the value of planning and controlling time use. If you need to fine-tune your system, review tips in Figure 3.7.

"Because I was a husband and a father of two young daughters, a full time engineering student who also worked part-time, and a great volunteer in my IEEE student branch, time management (or crisis management, in my case) was very important in my day to day activities."

(Sedofia Gedzeh, Electrical Engineer)

TOP TEN TIPS TO MANAGE TIME

1. Create an environment that minimizes distractions
2. Be realistic in identifying activities and establishing target dates
3. Break large activities into smaller steps
4. Focus on one activity at a time until it is completed
5. Schedule work breaks to rest and reenergize
6. Review work plan each day and revise as needed
7. Eliminate old habits that waste time
8. Schedule times for some "fun" activities
9. Learn to say no when appropriate
10. Delegate work to others

Figure 3.7
Top ten tips for more effective time management.

3.6.3 Enjoy More Free Time

Your reward in using an effective system is realizing you can accomplish important activities and deliver results on time or even early. You can reduce stress by preventing "all nighters" and last minute cramming for tests. In addition, you will be happy with your accomplishments and knowing you are in control.

Another value of an effective time management program is that it generates more time for personal activities. It is easy to get bogged down and miss some of the pleasures of life. By using time management, you will have more time to participate in many different activities in addition to school or work.

3.6.4 Develop Skills for Working Career

As a student, you are in an environment where you have many opportunities to practice and enhance nontechnical skills before starting your working career. Time management is one of the most useful of your nontechnical skills, so develop a good system now.

When you near graduation, you will be involved in the job search process. Companies are more competitive and successful when engineers complete work on schedule. So, interviewers are interested in hiring candidates who can produce results by using time effectively. Identify time management as one of your skills on your resume. Then, during job interviews, discuss how you complete projects on schedule and you will have a better chance to be considered for the job.

Time management is even more important in your working career.

When you are a good manager of time on the job, you will expand your value to the company. Although your assigned activities will change as you progress in your career, the principles of time management apply to all working environments. When you know how to plan your work, you will be ready for more comprehensive challenges throughout your career.

3.7 Time Management Impact on Career and Quality of Life

The common theme in all your activities is that they require the use of time. The more efficiently you use time, the more time you will have for expanded activities. The control of time use also brings the reward of accomplishments with less stress.

As you learn to work with instructors and other students, you can expand the quality of your life by learning new ideas and understanding different perspectives

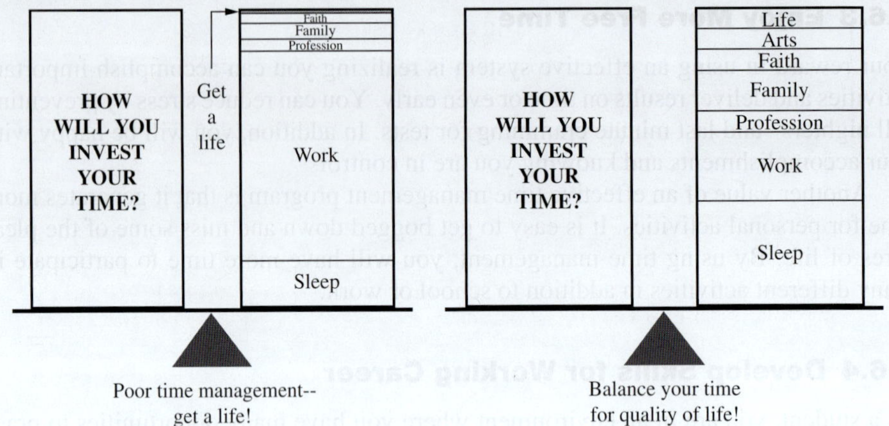

Figure 3.8
Time investment options.

from others. And your quality of life includes your involvement with your family and friends.

If you are not convinced that time management is worth the effort, perhaps the two diagrams in Figure 3.8 will cause you to give this greater consideration. One of the best ways to achieve high levels of success is to balance your professional and personal time to include the most important aspects of your life.

> ***"Budget your time like you budget your money. Have a plan for how you are going to spend all 24 hours of the day, even if that means you pencil in 'sit around and do nothing' as part of your day. Keep just one calendar that holds information about upcoming assignments, tests and personal commitments to avoid conflicts. The word 'no' is a powerful tool we often forget. Just because you are asked to do something doesn't mean you are obligated to do so. Be honest with yourself and those who ask you about the time you have to commit."***
>
> **(Kristi Brooks, BS Electrical and Computer Engineering, and Electrical Engineer at Ideal Aerosmith)**

When you feel you just don't have enough time to do the things you want to do, pause a moment and consider the message of the glass jar and two cups of coffee in Figure 3.9. You can always find time for the most important things in life if you keep things in perspective. You really do have time for two cups of coffee with a friend.

THE GLASS JAR

A professor filled a large empty glass jar with golf balls, and asked students if the jar was full. They agreed that it was.

Then, he poured pebbles into the jar. He shook the jar and the pebbles filled the open spaces between the golf balls. He asked the same question, and students agreed it was now full.

Next, he poured a box of sand into the jar and it filled the remaining empty spaces. He asked students if it was now full, and they agreed that it was.

To the amazement of students, the professor emptied two cups of coffee into the jar effectively filling the spaces between the sand.

"Now," said the professor, "This jar represents your life."

The golf balls are the big, important things–your religious faith, family, children, health, friends, and your passions–things that make your life full even if everything else is lost.

The pebbles are important things like your job, house, and car. The sand is everything else–the small stuff in your life.

If you fill the jar with sand first, there is no room for pebbles or golf balls. So, if you spend all your time on the small stuff, you will never have time for the more important things.

Therefore, manage your time and take care of the things that really matter first. Focus on the golf balls and pebbles and not the sand in your life.

A student asked what the coffee represented. The professor replied, "No matter how full your life may appear to be, you always have time for two cups of coffee with a friend."

Figure 3.9
The glass jar.

Enhance your quality of life by managing time.

Without planning and controlling your time, you will develop many poor habits that will definitely have a negative impact on your level of success. More importantly, your quality of life will be much less than you deserve. So, reduce stress by balancing academics, work, and professional involvement and then you will have time for other important life activities.

Your educational experience offers many opportunities to develop time management skills and these skills will improve your success as a student. Perhaps even more importantly, a successful time management system can be adapted to your working career and will help you achieve high levels of success while enjoying a balanced life.

A Conversation with Courtney Gras

Courtney, tell me how you established a time management system to handle all your activities discussed in the scenario at the beginning of this chapter.

Answer

"I've learned that defining an effective time management system comes from personal experience. Any professor will tell you time management is important, but if you are the type of individual who likes to be involved in numerous activities like I am, it becomes tempting to continue to add activities until something begins to suffer. Whether it is grades, relationships, or overall stress level, the first step is recognizing how much is too much. Once you have gained that experience it is easier to prioritize tasks, keep track of what needs to be done first, and filter additional tasks that can be omitted."

What are some of the nontechnical skills you use to manage time?

Answer

"Many of the nontechnical skills I used for time management come from obvious sources such as calendars (electronic or printed), automatically generated email reminders, and "to-do" lists (electronic or written). I discovered through an unpleasant experience that it is not a good idea to rely on technology to keep track of your life. Hard drives fail, data is lost or corrupted, and phones are broken. It is a good idea to have multiple ways of keeping track of what needs to be done, and keeping track and prioritizing by date is usually the best method.

Another nontechnical skill in time management is judgment. For instance, opportunities for college-related travel may present themselves in the form of conferences, competitions, etc. Participation may require missed classes, labs, or tests. A decision must be made on the overall importance and relation of this opportunity to the ultimate career goal before proceeding. If it is a once-in-a-lifetime opportunity, I use judgment to choose how my time should be spent."

What are the benefits of managing your time?

Answer

"By choosing to participate in extracurricular technical and nontechnical projects, I have gained knowledge, experience, expertise, and beneficial contacts. Since I employed proper time-management methods and prioritized my tasks, not only did I build my resume and have extremely enriching experiences that the majority of my classmates missed out on, but I also maintained my grades and at least some of my social life."

What advice would you share with students?

Answer

"Always take opportunities and go the extra mile—but plan your steps before you take them. If you employ time management methods and good judgment while knowing your limits, the success and rewards you will experience will likely go beyond anything you would have imagined. This will benefit your career in college, and prepare you for your future career."

(Courtney Gras, IEEE Student Chapter President, Robotics Team Leader, and Undergraduate Student at the University of Akron)

3.8 Conclusion

As Courtney has clearly stated, time management is one of the most important tools for being a successful student. Effective management of various activities leads to more accurate work, fewer time pressures, and a more satisfying life.

Time, to a great extent, is under your personal control. You select most activities and place priority based on your individual desires. Time management can provide a sizable expansion of time to be used for greater success in school and for other activities.

You can use many different approaches to time management. Consider several processes until you find one that feels comfortable. Select a system and then incorporate the proven principles of managing your time discussed in this chapter. You will like the results.

END OF CHAPTER REVIEW QUIZ

Select the most appropriate answer to the following statements.

1. One of the best ways to be more successful in school is to
- **a.** Work a lot of homework problems.
- **b.** Use a system to manage time and complete activities when due.
- **c.** Be active in a professional society.
- **d.** Learn how to communicate with others.

2. The first step in managing time is to
   a. Assign priority of activities.
   b. Use a calendar and put activities on due dates.
   c. Make a "TO" list of activities and due dates for each.
   d. Set aside a specific time each day to work on activities.
3. The most important value of a "TO" list is to
   a. Define what you need to do.
   b. Establish a start time for each activity.
   c. Show others how much you can do.
   d. Record your achievements for use in your resume.
4. The second step in managing time is to
   a. Assign a level of priority for each activity.
   b. Order the list by due date and priority.
   c. Select a start date.
   d. Enter data in an Excel file.
5. The third step in managing time is use a "DO"list to
   a. Document the results of completed activities.
   b. Revise your list each day.
   c. Start and finish activities.
   d. Determine how much time each activity needs and establish start dates.
6. Time management is most successful when you
   a. Start with easy activities to build confidence.
   b. Finish one activity before starting a second one.
   c. Document the results of each activity.
   d. Schedule most activities to be completed on weekends.
7. Students who are successful time managers
   a. Adapt a time management system that fits their lifestyle.
   b. Revise their "TO-DO" list on a daily basis.
   c. Break large activities into smaller activities.
   d. Do all of the above.
8. The primary reason some students stop using time management is
   a. They learn how to get things done without a time management system.
   b. They get discouraged when they see how many things they have to do.
   c. They try to use a system that is too complicated and time consuming.
   d. They don't like to be organized and complete activities on an emergency basis.
9. The best way to make time management more effective is to
   a. Select the most appropriate system for your needs.
   b. Use proven methods to identify activities, select start dates, complete work, and then have time for leisure and recreation.
   c. Follow your plan and do not procrastinate.
   d. Do all of the above.
10. The greatest value of time management is
    a. To achieve a good grade in this class.
    b. To be organized so you do not waste time.
    c. To successfully complete all activities and have more time for things you enjoy.
    d. To learn how to use a time management system.

EXERCISES TO DEVELOP AND ENHANCE YOUR SKILL SET

Exercise 3.1 This is an exercise to help you develop a workable time management system. Use a blank 8′ × 11′ sheet of paper and write the following column headings:

ACTIVITY DUE DATE ABC HOURS START DATE

Practice using the TO-DO Process by completing the following steps:

1. List six activities that you need to complete in the next two weeks.
2. Write the due date for each of these activities.
3. Assign A, B, or C to identify the level of importance for each activity.
4. Estimate the amount of time to complete each activity and list this in the "HOURS" column.
5. Select the start date for each activity by considering the due date, amount of time to complete, and the level of importance.

Use the results of this exercise as a foundation for the rest of the exercises for this chapter.

Exercise 3.2 This assignment is an opportunity for you to focus on a successful time management process. If you have been using a time management system in the past, use this exercise to evaluate how effective it is in helping you invest your time.

If you have not been using a structured time management system, now is a good time to review options and select a trial system. You can develop a form similar to Figure 3.10, "Time Management Excel Worksheet," or select a system of your choice for this assignment.

Time Management Excel Worksheet				
Activities	Due Date	ABC	Hours to Complete	Start Date

Figure 3.10
(Also available from the Web as an Excel worksheet.)

Although there are many tools and different support systems to help you, the most important part of this exercise is to see how the basic principles of time management discussed in this chapter apply to your system.

Your assignment is in three (3) parts.

Part 1—Use a time management system for two weeks

- Use a process that identifies activities, assigns due dates, level of priority, and start dates.
- Identify activities to be completed during the time assigned by your instructor.
- Break large and long-term activities into manageable subactivities.
- Complete activities as scheduled.
- Document and evaluate results.

Part 2—Prepare a written report

- Prepare a written report using the format shown in the Time Management Report form on the next page.
- Include your name, suggested title, and subtitles, as shown on the format.
- Review comments in each section for guidance of requested content.
- Use good English grammar, punctuation, and sentence and paragraph structure.
- Select 11 or 12 point font, and use left margin justification.
- Use spellchecking and grammar software, print a draft, review, and make corrections.
- If assigned by your instructor, use the resources of University Writing Center to review your report and follow their advice for writing improvements.

Part 3—Submit your report for evaluation

- Submit a printed copy of your report and attachment at the start of class on the date assigned by your instructor.
- Attach a printed sample of your time management system.
    - Use the form shown in Figure 3.10 (also available on the Web as an Excel worksheet) or your existing system format.
    - Print a sample showing activities for the two weeks in this assignment.
- Send an electronic copy of your report to your instructor as assigned.

Time Management Report

Prepared by ____________________
(Student name)

Time Management System

Describe the system used to manage your activities during this two-week period. Explain if this was a new system or one that you have been using. This should be one or two *paragraphs.*

Top 6 Most Important and Valuable (A) Activities

Use *bullets* to list the top 6 most important and valuable (A) activities. Include due dates and dates on which each activity was started and completed.

Summary of Top 6 Most Important and Valuable (A) Activities Results

Use *paragraphs* to report significant information relating to the results of planning and completing these activities. This does not need to be lengthy, but should describe how your time management system worked for the most important and valuable activities.

Top 6 Important (B) Activities

Use *bullets* to list the top 6 important (B) activities. Include due dates and dates on which each activity was started and completed.

Summary of Top 6 Important (B) Activities Results

Use *paragraphs* to report significant information relating to the results of planning and completing these activities. Discuss the difference in how you worked with these activities (B) compared to those that were most important and valuable (A).

Time Management System Assessment

Write one or two *paragraphs* to report your overall assessment of your time management system and this exercise. Rate your experience from 1 to 10 (10 being the highest value) and give reasons for your numerical rating.

Discuss how you will apply what you learned in this experience to your future time management process.

CHAPTER

4 Communication Skills

Information Transfer
Developing Effective Communication Skills

"Your success as an engineer will be directly proportional to your ability to communicate!"

—Dr. Charles K. Alexander,
Author of *Fundamentals of Electric Circuits*

COMMUNICATION SCENARIO

"A few years ago, when I was a policy fellow in Washington, DC, I received a call at 8:30 AM *from a colleague who gave me a 'top priority' project: prepare a one-page policy paper that would be used to brief a congressman during a ten-minute car ride from his Capitol Hill office to a noon meeting at the White House. The brief was to focus on policy challenges in global neglected diseases. 'The brief is due at 10:00* AM*,' my colleague repeated and abruptly hung up. Thirty panicked seconds later, I had Google open on my browser and was (over)confident that I would be able to accomplish the task."*

(Guruprasad Madhavan, Biomedical Engineer and Program Officer, National Academy of Sciences)

This scenario, prepared by Guruprasad (Guru) Madhavan, demonstrates how important it is to develop good communication skills early in your career. After we discuss the fundamentals of communication, we will discover the rest of Guru's story at the end of this chapter.

4.1 Introduction

Do you think all you need to be a successful engineer is the ability to learn and apply technology? If your answer is yes, the following paradox may change your mind. Engineers must have knowledge and skills to apply technology, but success is directly proportional to their ability to apply nontechnical skills. If your answer is no, then you are already aware of the importance of nontechnical skills.

LEARNING OBJECTIVES

By using the information and exercises in this chapter you will be able to

- Understand why communication skills are so important to your career success.
- Increase comprehension by improving your reading skills.
- Organize and structure written documents.
- Develop and use professional writing skills to establish your credibility.

A review of the Table of Contents makes it clear that this entire book is based on improving your nontechnical skills. We discuss one of the most important of these skills in this chapter, and that is communication.

> *"Communication is a learned skill. Most people are born with the physical ability to talk, but we must learn to speak well and communicate effectively. Speaking, listening, and our ability to understand verbal and nonverbal meanings are skills we develop in various ways. We learn basic communication skills by observing other people and modeling our behaviors based on what we see. We also are taught some communication skills directly through education, and by practicing those skills and having them evaluated."*
>
> (Wikipedia)

We start with an overview of the basic fundamentals of communication you will need to be successful. Unlike some resources that only discuss communication, this chapter includes exercises to help you learn how to use engineering tools to enhance your communication skills. When you are a good communicator, you will reach higher levels of success in school and in your working career.

4.2 Why Effective Communication Is Important to You

Think about it. What value is technical knowledge if you can't share it with others in a practical way so they can understand and use it?

When you are an engineer, you may sit in a cubicle with your computer or in a laboratory and discover important ideas and information. Unless you take the next step and communicate what you know to other people, nothing will change. You can feel good about what you know, but how does that make a difference in the world? Your knowledge becomes valuable only when you transmit it to others.

> *"Effective communication is an important skill set that is often overlooked. The technical ability to understand problems and provide solutions is the backbone of a company's core capability to compete in today's challenging business environment. Engineers who are able to explain their ideas and solutions in both written and verbal formats are generally more respected by their peers and competitors. This additional recognition provides more opportunities to choose your assignments and control your fate."*
>
> (Ted Tracy, Electrical Engineer and Regional Manager)

4.3 The Communication Process

Various ways to communicate follow a similar pattern:

- One person (sender) has information (message) to transmit to other people (receivers).
- The sender decides how the message will be transmitted.
- The message is sent.
- The receiver hears or reads the message.
- The receiver processes the message.
- The receiver may or may not give feedback to the sender.

4.4 The Four Basic Types of Communication

The four basic types of communication are reading, writing, listening, and speaking. In this chapter, we discuss the value of reading and writing and their importance as communication skills for students and practicing engineers. Listening and speaking are covered in other chapters.

> *"My recommendation is to master the 'soft skills' of reading, writing, listening, and speaking, since these are absolutely needed to further one's career. As you advance, you will be communicating with people who have less and less of a technical background, so know your audience and direct your communications to the level of the audience."*
>
> (Howard Wolfman, PE, Electrical Engineer and Principal, Lumispec Consulting)

You may be surprised to learn that we are involved in reading and writing for only about one-fourth of our communication activities, as indicated in Figure 4.1. However, our ability to read and write effectively has a major impact on our ability to communicate.

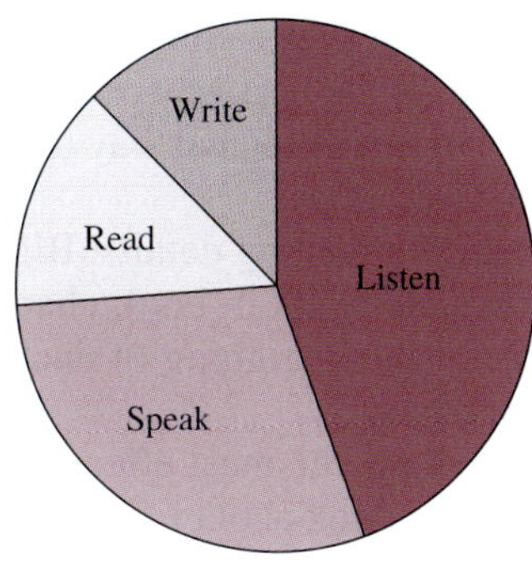

Figure 4.1
Typical communication time use.

4.5 Reading

First, let's put reading in perspective. As you consider the many electronic vehicles for delivering information, you might be tempted to think printed information on paper or in books is becoming obsolete. After all, you can search everything on the Internet, right?

Although there are new and different ways to obtain information, printed information will be used for many years. It may be on paper or in an electronic book or on another device, but printed information will be available in some form for a long time.

You have different reasons to read printed information. When you read for pleasure, you can relax with little concentration and just be entertained. When you read to gain important information related to your education and work, it requires more effort on your part. So, the way you read depends on what you are reading and what you want to gain from doing so.

4.5.1 Speed Reading

You may be acquainted with an interesting method of reading called speed reading. This method is often used to read section subtitles and the first sentence of each paragraph as a fast way to review material. The first sentence usually identifies the paragraph topic and can give an outline of the overall information. So, speed reading can be used to obtain an overview of information to identify areas for more detailed review.

4.5.2 Casual Reading

Casual reading is used to gain a general understanding of material. Because you want to understand and retain some of the information when involved in casual reading, your reading pace is slower than when speed reading. This slower pace makes it possible to think about information and how it relates to other topics. Casual reading is often used in reading fiction and less technical materials.

4.5.3 Comprehensive Reading

The third reading style is comprehensive reading, and this takes more concentration so your pace is much slower. This is the best way to obtain information when reading textbooks and other technical material. A logical way to use comprehensive reading is to break the process into a few basic steps.

Start with the big picture before reading details. Big pictures include the book's table of contents, introduction statements in book chapters, and report summaries. A good example is the outline at the beginning of each chapter of this book under the heading of **Learning Objectives.**

After you identify the big picture, the most effective way to understand information is to read at a slow pace, and to think about information and how it relates to your current knowledge. Comprehensive reading results in a greater level of retention.

Reading technical books is much like listening to class lectures. In class, you take notes of the most important concepts to help you understand and remember them. You can use a similar approach when reading books to learn and understand important information.

Book notes, like lecture notes, enhance comprehension.

As demonstrated by the student making notes while reading a textbook in Figure 4.2, you can enhance your learning process when reading textbooks by writing notes about major ideas and identifying areas that need further explanation. Use your notes to ask

Figure 4.2
Writing notes during comprehensive reading.

questions during lectures on the same topics. By the way, asking questions in class is an excellent way to expand your ability to understand and use information from your textbooks.

Reading skills are important because you need to keep up with the rapid change in technology. You will be more successful when you develop the ability now, as a student, to read printed and electronic information with good comprehension.

4.6 Writing

"As an engineering student, I know written communication is an integral skill that is responsible for a successful transfer of information, a stimulating exchange of ideas, and the development of a professionally enriching network. After all, no matter how profitable or unique a new idea may seem to be, without the open and effective streams of communication, it will be impossible to convey its importance and significance to others."

(Vladislava Cuznetova, student at Cleveland State University)

When you are given an assignment to write something, how do you feel? Do you think writing is a difficult task? Do you have a "mental block" and find it hard to get started. If writing is a challenge for you, I have good news. You can apply the principles of project management, as outlined next, to "engineer" your writing tasks and become a good writer and communicator.

PROJECT MANAGEMENT PROCESS

1. Preplanning
2. Planning
3. Organizing
4. Modeling
5. Completing project

Project management details are discussed in another chapter. However, you can start developing your project management skills now by applying them to writing. Let's use a typical laboratory report as an example of how project management steps can be applied.

4.6.1 Preplanning

Start with the big picture.

You might be surprised that there is a step before planning. Successful project managers do not get involved in project details until they define the big picture. This is called preplanning because it is based on developing an overall perspective before designing the project.

To see the big picture in a writing project, start by answering the following questions:

- Who will read what I write?
- Why would they be interested in my information?
- How will they use this information?
- How much information do they need?
- What writing style is best for them?
- What is the deadline to complete this writing project?

Preplanning is primarily identifying your readers and determining why they will read your material. By focusing on the reader, you gain a better idea of what to include and how much detail is appropriate. Then, you can use the proper tone and structure the content to be most effective for the reader.

HEAT TREATMENT AND HARDNESS

The objectives of this experiment are to become familiar with metallographic practices and to learn how to analyze mechanical properties. This was done by examining the correlation of heat treating steel and the resulting properties of hardness and strength.

Figure 4.3
Introduction of laboratory report.

Figure 4.3 is an example of a student's introduction to a laboratory report. The reader is the instructor, so the introduction outlines the objective of the experiment and gives an overview of what was done. The instructor is interested in the report because it was an assignment and the report evaluation results will be applied to the student's grade for the course.

The amount of information should be sufficient to describe the experiment in a clear and concise way to help the instructor evaluate the report. The writing style is somewhat formal and follows the format provided by the instructor. The deadline is established in the assignment.

4.6.2 Planning

After answering preplanning questions, the next step is to make an outline. A good approach is to use word processing software and identify major sections. Select information to fit the objective, as determined in the preplanning step. Most readers are busy, so they need to know enough facts to comprehend your message, but not so much that it becomes "information overload."

As you consider what is important to readers, you can expand the content and increase your credibility by researching various sources of information relating to your topic. When you reference information from others, you demonstrate that you have researched a variety of sources and included more than your own thoughts.

So, where do you start your research? Your first thought may be the Internet. This can be a good source of information as long as you recognize the source may, or may not, be credible. The Internet can save you time by identifying where information can be found.

University and public library reference sections are excellent sources of information. In addition, librarians can save you time by directing your search to appropriate sources. During your search, consider the credibility of the source as well as the information.

Textbooks, technical journals, and published articles typically are reviewed by peers, so they are based on facts and should be reliable sources. Most professional societies have electronic and printed libraries and you can find information relating to articles on their websites.

As you collect information, record the source so this can be added to your reference list as you develop your report. Reference information includes the author, title or content description, and publication. A correct format for references is shown in the following examples:

Smith, J. M., and Van Ness, Hendrick, *Introduction to Chemical Engineering Thermodynamics,* 7th ed. New York: McGraw-Hill, 2005.

Alexander, Charles K., and Sadiku, Matthew N. O., *Fundamentals of Electric Circuits,* 4th ed. New York: McGraw-Hill, 2009.

Reference information is placed at the end of written reports in a Bibliography Section just before the Index.

4.6.3 Organizing

The third step, organizing, is one of the most important. Engineers typically use logical thinking to solve problems, and a logical approach makes it easy to organize written material. In addition, the use of computers adds a very efficient way to prepare and organize an outline.

Using outline identification symbols will help you place information in the correct area of a well-organized document. Outlines save time, because the structure is established before details are added. Two typical outline formats are shown in Figure 4.4.

	Roman Numerals, Letters, Numbers	Decimals		
Major parts	I, II, III	1	2	3
1st subparts	A, B, C	1.1	2.1	3.1
2nd subparts	1, 2, 3	1.1.1	2.1.1	3.1.1
3rd subparts	a, b, c	1.1.1.1	2.1.1.1	3.1.1.1

Figure 4.4
Typical outline formats.

LABORATORY REPORT OUTLINE

I. Introduction

- **A.** Objective
- **B.** Method of Application
- **C.** Summary of Results

II. Experimental Procedure

- **A.** Equipment and Use
- **B.** Outline of Experiment Steps

III. Discussion

- **A.** Data summary
- **B.** Name and Values of Variables
- **C.** Analysis Process
- **D.** Errors
  - **1.** Physical Phenomena
  - **2.** Instrument Limitations

IV. Conclusions

- **A.** Results
- **B.** Analysis
- **C.** Recommendations

Figure 4.5
Sample outline.

A good example of an outline structure is shown for the sample laboratory report in Figure 4.5 Notice that the first step in organizing is to identify the major parts to be included and their location in the report. Unless you are given a specific format, think about your readers and place information in a logical flow so it is easy for them to follow and understand.

4.6.4 Modeling

An important step in building a physical product is to make a model, determine if it needs to be changed or refined, and make appropriate corrections before completing the final product. A similar approach can be used in developing your written document.

The outline is a model and provides a roadmap by defining where to place information. Add details in each section of the outline. Consider how much and what information is appropriate, based on the preplanning results. When you have completed all sections, you have a working model to review and revise.

An example of how details are added to the laboratory report outline in Section I, Subsection C, Summary of Results, is shown in Figure 4.6.

LABORATORY REPORT SAMPLE

I. Introduction

C. Summary of Results

When carbon steel was heated to a temperature above 1,400° F, the structure was altered and an iron phase change occurred. The rate of cooling affected the final microstructure of the steel.

If cooling is slow, the iron carbide is forced out of solution and the gamma iron returns to alpha iron. The resulting formation consists of a series of plates of iron carbide interspersed with plates of ferrite. This is known as pearlite, and slow cooling produces coarse pearlite.

Faster cooling rates cause closer spacing of plates and the result is medium or fine pearlite, depending on the rate of cooling. Hardness increases from coarse to fine pearlite.

Figure 4.6
Information in a typical section of the laboratory report.

Up to this point, we have been discussing how to select and organize ideas into a written format. When you have included all appropriate information, the next step is to fine-tune the document. This can be done by using spelling and grammar checking software and by reading drafts.

Spelling and Grammar While content is important, you also need to consider how your written work is packaged. You can establish a professional image by using correct writing principles and an interesting writing style. These will eliminate barriers that could distract readers. Correct English grammar, such as subject/verb agreement and proper punctuation, is the foundation for good writing.

Start by running spelling- and grammar-checking software to identify misspelled words, errors in punctuation, subject/verb agreement, and other grammatical errors. Make corrections and then read your work on the computer screen. Test your model by reviewing the structure, and if necessary, rearrange information for better continuity. Think about how your readers will use information and make it easy for them to follow.

Word Selection As indicated in Figure 4.7, software checks spelling but not word selection. Figure 4.8 shows the corrected version.

I all ways used my spell checker to bee sure there are know errors. It's alright two take advantage of these resource.

You're reports will help your go farther than you ever though you could bye spelling all works properly.

Figure 4.7
Incorrect word selections.

I always use my spell-checker to be sure there are no errors. It's all right to take advantage of this resource.

Your reports will help you go further than you ever thought you could by spelling all words properly.

Figure 4.8
Correct word selections.

FREQUENTLY MISUSED WORDS

WORD	MEANING
Already	Completed earlier
All ready	Completely prepared
Amount	Sum or total
Number	Units that can be counted
Can	Have ability or power to do something
May	Have permission to do something
Continually	Repeated, interrupted action
Continuously	Uninterrupted action
Effect	Noun – result
Effect	Verb – to make happen
Affect	To influence
Farther	Distance
Further	Time or quantity
Imply	To suggest or indicate
Infer	To deduce from evidence
Lie	To recline
Lay	To put or place
That	Defines or restricts (one of two or more)
Which	Additional information without defining or restricting

Figure 4.9
Sample of frequently misused words.

It is easy to use words incorrectly if you do not understand their precise meaning. Technical terms are typically not a problem. But many common words are used improperly and can create barriers. If readers have to stop and think about what a word means, they can become distracted from the content of your message.

You can improve word selection by understanding the meaning of words. An easy way to start is to review commonly misused words in Figure 4.9.

Sentence Structure Improve your writing style by using interesting sentence and paragraph structures. For most technical writing, sentences should be short and written with a variety of simple, compound, and complex sentence structures.

Use a variety of sentence structures.

Good writing style increases reader interest and comprehension. Use a sentence to express a complete thought. To be an effective writer, use a mix of simple, compound, and complex sentences to add variety. We can demonstrate different sentence types with a few examples.

Simple sentences contain one subject and one predicate. A sample of a simple sentence is

New computer software is user-friendly.

In contrast, a compound sentence expresses two or more independent but related thoughts of equal importance. These are joined by a comma followed by or, and, or but. Compound sentences are a merger of two or more simple sentences or independent clauses that deal with the same basic topic. A sample of a compound sentence is

New computer software is user-friendly, and it is much easier to install on most personal computers.

A complex sentence expresses one independent thought and one or more subordinate thoughts (dependent clause) that relate to it, separated by a comma. The subordinate thought cannot stand alone and typically is placed at the start of the sentence. A sample complex sentence is

Although computer software changes frequently, recent software is user-friendly.

Selection of the type of sentence structure should match the relationship of ideas. If you have two ideas of equal importance, they should be expressed as two simple sentences or in one compound sentence. However, if one of the ideas is less important, it can be placed as a dependent clause in a complex sentence.

Long sentences are usually more difficult to understand than shorter ones because the reader has to work harder to follow the large amount of information. So, it is better to break long sentences into two or more sentences. Articles written in the magazine, *Reader's Digest,* are excellent examples of simple, but effective, sentence structures.

Paragraph Structure Proper paragraph structure also impacts reader comprehension. A paragraph should cover only one main idea, and this idea should be introduced in a topic sentence.

A paragraph covers one main idea.

Typically, a topic sentence is the first one in the paragraph; however, it can also be included within the paragraph or even as the last sentence. The balance of the paragraph should include one or more sentences to support and expand the main idea. If sentences do not support or relate to the main topic, they should be placed in a new paragraph.

As in the case of sentences, shorter paragraphs are usually easier to follow and understand. The use of concise statements and a limited number of supporting sentences will result in paragraphs of reasonable lengths.

Transition Statements Transition statements help readers move from one thought to the next. Transitional words, phrases, and sentences are tools for readers to see how different ideas relate to each other either within a paragraph or between two paragraphs.

Use transition statements to link paragraphs.

You can use transition words and transition sentences to build continuity. These connections can be established in the last sentence of a paragraph to link to the

RELATIONSHIP	TRANSITION WORD EXAMPLES
Comparison	in comparison, still, similarly, likewise
Contrast	conversely, whereas, nevertheless, however, but
Repetition	that is, in other words, as has been stated
Illustration	for example, in particular, in this case, for instance
Additional detail	furthermore, in addition, besides, first, second, finally
Relationship	therefore, because, accordingly, thus, consequently, so
Time sequence	formerly, after, when, meanwhile, sometimes
Summary	in summary, to sum up, in brief, to conclude, in conclusion

Figure 4.10
Transition statements.

following paragraph, or they can be included in the first sentence of the next paragraph. Sample transition statements are shown in Figure 4.10.

Transition examples are shown in bold in the following paragraphs:

EXAMPLE 4.1

The purpose of this research project is to increase system efficiency of existing technology. Using variations of fluid piping designs increased the efficiency of the hydraulic system from point A to point B by 10%. **Before recommending the use of these design changes, one additional comparison should be made.**

The initial cost for the most efficient design was 15% higher than previous designs. This may be acceptable if operating cost savings offset the first cost within 3 years.

EXAMPLE 4.2

However, if initial cost is the most important factor for consideration, it may be more acceptable to consider the application of new technology. Some systems, using synthetic fluids, have entered the market with relatively low first costs.

EXAMPLE 4.3

In summary, the original purpose of this study may be too narrow to produce practical results. Additional studies are recommended.

4.6.5 Completing the Project

"The most valuable of all talents is that of never using two words when one will do."

(Thomas Jefferson, third president of the United States of America)

To complete your writing project, print a draft and review the overall structure by reading it aloud. Are topics and subtopics appropriate? Are various topic areas related, and do they fit into the overall story? Is the content written concisely without extra, unneeded words?

Prevent barriers by using common words and eliminating unnecessary ones.

As a result of the fact that four of our astute committee members could not be present at our last scheduled meeting time on today, June 25 at 9:00 AM, the liberty was taken to cancel that assemblage. It is, therefore, planned to consider the same agenda items at the newly scheduled aggregation, which will convene the forthcoming Friday morning, June 27 at 9 AM. Therefore, it is my personal decision that this same committee meeting will be convened in accordance with the previously established agenda which I have repeated herein for your convenience to review before the meeting and to help you be better prepared to discuss agenda items in significant detail at the meeting on Friday.

Figure 4.11
Wordy message.

Because four members could not attend our last meeting, I rescheduled it to 9 AM next Friday. We will use the same attached agenda.

Figure 4.12
Concise message.

Think about readers as you review your final draft. Does your message create a barrier like the wordy example in Figure 4.11, or is it concise and easy to understand like Figure 4.12? Are the length and variety of sentences appropriate to generate reader interest? Are paragraphs easy to read? Did you use appropriate transitions to enhance the flow of information?

The final step is to make corrections from the last review, use your spell-checker one more time, and print the final product. If you have worked hard and followed each step of the draft process, you will have a very professional and effective written document.

A CONVERSATION WITH GURUPRASAD MADHAVAN

Guru, the scenario you described in the beginning of this chapter is a little overwhelming. How did you handle this challenge to prepare something so important so quickly?

Answer

"My work in public policy not only gives me the opportunity to look at issues differently, but also challenges me to communicate better.

In this situation, as I was looking at the Internet, I was blindsided with hundreds of thousands of results. After 45 minutes of research online for credible information—mind it I hadn't written a word for the brief yet—I realized I was pedaling at the same location without moving forward. I detached myself from the computer, changed my research strategy, and started calling experts in this area for guidance. Their suggestions helped me to frame the topic and start writing. Alas, the time was already 9:45 am. I felt I needed at least an extra hour to chisel the issues.

I called my colleague and said, 'I blew it! I don't think I'll be able to finish the brief by 10:00.' She chuckled and revealed that this was a training exercise to help me get adapted to the realities of the business as part of my 'startup' package in public policy.

My colleague and I later had coffee together to go over what I had developed during the exercise. I learned a lot about improving both my writing and storytelling skills under pressure."

Although this turned out to be a test run for you, how do you typically prepare for briefings? What nontechnical skills do you use?

Answer

"If I hadn't started making phone calls to experts, I'd have woefully failed in my research. I had to achieve an escape velocity to overcome my fear and shyness in calling world leaders in the subject. Engineers can learn a great deal from talented journalists who are renowned for asking thoughtful questions to elicit thoughtful answers to help build a thoughtful story.

Engineers, in general, are competent in tackling complex data and models, but that skill alone is not sufficient in an overwhelmed society with ultra-short attention spans. More impressive to me than data-dense reports filled with charts, graphs, and numbers is a story line that's well thought out, buttressed with relevant data, and presented in an accessible and engaging style. Always remember: Stories stick.

This experience has bolstered my interests in telling stories supported by solid research findings. It is these factually accurate stories, coupled with objective persuasive skills that help create actionable policies."

What advice would you share with students?

Answer

"Read, for example, *The Economist, The New Yorker, Foreign Policy, Scientific American,* and other magazines, books, and guides. Search for talented writers whose working style will inspire you to sharpen your research, writing, and storytelling skills.

Also focus on reading broadly beyond your subject area, including areas in the arts, design, business, health, finance, economics, law, psychology, sociology, humanities, policy, history, government, and international development. As 21st-century engineers, our challenges are far more crosscutting and complex than they have been in the past."

(Guruprasad Madhavan, Biomedical Engineer and Program Officer, National Academy of Sciences)

4.7 Conclusion

We have discussed two of the four communication processes, reading and writing, in this chapter on communication. Other chapters will cover listening and speaking. All four communication processes are very important to help us receive and deliver information.

We also discussed how we use different reading styles when reviewing printed and electronic messages. These included speed reading for a quick overview, casual reading for pleasure, and comprehensive reading when we need to understand and use information.

In our discussion of written communication, we identified the most important principles of keeping a focus on the reader, using good organization, and providing documentation to add credibility to our technical documents. Writing should be based on proper English grammar and professional writing techniques. Written documents can last a long time, so we should prepare them in a correct and professional way to establish a positive image.

Although applying correct and concise writing skills takes effort and time, it results in many benefits. Professional writing will enhance your credibility and will encourage others to understand the value of your ideas. This, in turn, will open many doors of opportunity for advancement and increased career success.

"I have witnessed throughout my career, many brilliant engineers that border on the line of genius. Unfortunately, some of these sharpest minds were not considered to be promotable because of a deficiency in communication skills. As an engineering student, spend the extra time and effort on your lab reports and writing assignments. It will pay off in ways that at this point you can only imagine."

(Ted Tracy, Electrical Engineer and Regional Manager)

Ted Tracy's comments summarize the value of good communication skills. Many engineers fail to reach high levels of success, not because they lack technical skills, but because they cannot communicate with other people. So, take Ted's advice and use the many opportunities you have as a student to develop effective communication skills. This will pay big dividends throughout your life and career.

END OF CHAPTER REVIEW QUIZ

Select the most appropriate answer to the following statements.

1. The best way to prepare to be a successful engineer is to
- **a.** Study hard and earn good grades.
- **b.** Balance technical skills with nontechnical skills.
- **c.** Learn how to apply technical skills.
- **d.** Develop good writing skills.

2. The four basic types of communication are
- **a.** Reading, writing, listening, and speaking.
- **b.** Talking to friends, listening to speakers, text messaging, and talking on a cell phone.
- **c.** Writing reports, sending e-mails, preparing resumes, and sending cover letters.
- **d.** Turning in homework, asking questions in class, reading textbooks, and taking tests.

3. The major parts of the communication process are
- **a.** Learning how to write and speak.
- **b.** The three types of reading—speed, casual, and comprehensive.
- **c.** One person sends a message, others receive and process the message, and others may give feedback to the person sending the message.
- **d.** To take notes when reading textbooks, ask questions in class, and listen to answers.

4. Speed reading is an effective way to
- **a.** Quickly obtain an overview of written material.
- **b.** Study for a test.
- **c.** Read books for pleasure.
- **d.** Read the chapter in a textbook just before class to prepare for the lecture.

5. Comprehensive reading is a good way to
- **a.** Read the textbook before class.
- **b.** Prepare for a laboratory experiment.
- **c.** Study material in preparation for a test.
- **d.** Accomplish all of the above.

6. The best result of writing notes when reading a textbook is
- **a.** It makes sure you read the entire book.
- **b.** It will help you identify questions to ask in class for more information.
- **c.** It provides evidence you can use to show your instructor you read the book.
- **d.** To slow your reading speed and help you understand all the details of a text.

7. When starting a writing project, the first step is to
- **a.** Make an outline of the material.
- **b.** Write the first paragraph of each part of the written document.
- **c.** Write the first draft and use spell-checking software to correct errors.
- **d.** Identify readers, establish the writing objective, and decide what to write.

8. The most efficient method of preparing a formal report is to
- **a.** Start with preplanning activities to see the big picture.
- **b.** Make an outline and arrange information in a logical structure.
- **c.** Use drafts to review and revise information.
- **d.** Accomplish all of the above.

9. The best way to eliminate barriers when writing is to
- **a.** Use outline identification symbols so readers know where information is located.
- **b.** Use spell-checking software to eliminate misspelled words.
- **c.** Write concisely and use the correct tone for the reader.
- **d.** Use tables to show detailed information.

10. Learning to write well is important because it
- **a.** Delivers your information to others in a way they can understand and use it.
- **b.** Helps you communicate more effectively with people.
- **c.** Is evidence of your knowledge and leads to greater career success.
- **d.** Is all of the above.

EXERCISES TO DEVELOP AND ENHANCE YOUR SKILL SET

Exercise 4.1 This is a reading exercise. Evaluate comprehension by using three reading methods to review the excerpts from the Introduction section of the *Fundamentals of Electric Circuits* book, Chapter resource 4.1, as follows:

1. Speed Reading
   - Read the seven paragraphs in 30 seconds. (Hint: If you have not been trained in speed reading, look at the first sentence in each paragraph and quickly move down the page to obtain a visual image of content.)
   - Without looking at the excerpts again, write a summary of what you have just read.
2. Casual Reading
   - Read the paragraphs again in 5 minutes, but do not take notes as you read.
   - Without looking at the excerpts again, write a summary of what you have just read.
3. Comprehensive Reading
   - Read the paragraphs a third time in 5 minutes and take notes as you read.
   - Without looking at the excerpts again, use your reading notes and write a third summary.

Class discussion (or your study team if it is not discussed in class):

1. Compare the content of all three summaries.
   - When would speed reading be appropriate?
   - If you were preparing for a test, which method would you choose?
   - What method would you use if details are not important and you only need a general understanding of material?
2. Use notes from this exercise and class (or study team) discussion to complete the homework assignment.

Exercise 4.2 This is an exercise in developing paragraphs. Review the information shown below and determine where new paragraphs should start.

People, equipment, space, and energy are a company's resources. They are expensive, and you want to use them effectively. Productivity is a measure of resource use, and is the ratio of output over input. To increase productivity, you need to increase output, reduce input, or a combination of the two. The location of services like restrooms, locker rooms, cafeterias, tool cribs, and any other service will affect employee productivity and, therefore, the employees' utilization or effectiveness. It is said that you can run pipe and wire, but you cannot run people. Providing convenient locations for services will increase productivity. Equipment can be very expensive, and the operating costs must be recovered by charging each part produced on that machine a portion of the cost. The more parts run on one machine, the lower is the unit cost that each part must carry. So to achieve the second supporting objective, namely, to reduce cost, you must strive to get as much out of each machine as possible. Calculate how many machines are required in the beginning for maximum machine use. Remember, machine location, material flow, material handling, and workstation design all affect equipment usage. Space is also costly, thus designers need to promote effective use of space. Good workstation layout procedures will include everything required to operate that workstation, but not extra space. Normally, planners can do a good job of using floor space, but there are also other spaces for consideration. Basements are a good place for utility tunnels, walkways between buildings, under-the-floor conveyors for material delivery or trash removal, and tanks for storage. Overhead spaces may also be usable. These can be used for overhead conveyors, pallet racks, mezzanines, shelves or bins for storage, balcony offices, pneumatic delivery systems, dryers, ovens, and so on.

Exercise 4.3 This is an exercise in preparing a written report.

Prepare a written report based on notes from Exercise 4.1 and turn in your report at the beginning of class on a date assigned by your instructor. Prepare your report as follows:

Reading Exercise Report

Prepared by ______________________________

Application of Reading Methods

Speed Reading

Identify when speed reading would be appropriate.

Discuss the value of speed reading for your personal use.

Casual Reading

Identify typical uses of casual reading.

Discuss the value of casual reading for your personal use.

Comprehensive Reading

Identify when comprehensive reading would be appropriate.

Discuss the value of comprehensive reading for your personal use.

Evaluation of Reading Exercise

Write at least three paragraphs to compare results of each reading method as they relate to comprehension and retention of content.

Use a separate paragraph to discuss how you plan to use what you learned in this exercise in future classes and other reading experiences.

Exercise 4.4 This is an exercise to enhance your writing skills. Prepare a written report on the courses you are taking this semester and how you plan to learn new skills or enhance skills associated with each course.

Organize your report with section titles and a logical flow of information. Use proper sentence and paragraph structure and English grammar. Check spelling and grammar with appropriate software and correct errors before submitting report.

Your report should be one to two pages (400 to 800 words) in length.

Step 1:

Prepare your report using the process outlined in this chapter. Give an electronic copy of your report to your instructor no later than the established due date for this step.

Step 2:

If assigned by your instructor, visit the University Writing Center and request them to review your report and suggest how you can improve your writing skills. Work with their staff to revise and prepare a final report. Give an electronic copy of your final report to your instructor no later than the established due date for this step.

If your writing skills are evaluated with codes on your printed report, use Chapter resource 4.2, Writing skills reference, for an explanation of each code.

END OF CHAPTER RESOURCES

Chapter resource 4.1

Excerpts from the Introduction of the Fourth Edition of *Fundamentals of Electric Circuits,* by Dr. Charles K. Alexander and Dr. Matthew N. O. Sadiku:

> Electric circuit theory and electromagnetic theory are the two fundamental theories upon which all branches of electrical engineering are built. Many branches of electrical engineering, such as power, electric machines, control, electronics, communications, and instrumentation, are based on electric circuit theory. Therefore, the basic electric circuit theory course is the most important course for an electrical engineering student, and always an excellent starting point for a beginning student in electrical engineering education. Circuit theory is also valuable to students specializing in other branches of the physical sciences because circuits are a good model for the study of energy systems in general, and because of the applied mathematics, physics, and topology involved.
>
> In electrical engineering, you are often interested in communicating or transferring energy from one point to another. To do this requires an interconnection of electrical devices. Such interconnection is referred to as an *electric circuit,* and each component of the circuit is known as an *element.*
>
> A simple electric circuit consists of three basic elements; a battery, a lamp, and connecting wires. Such a simple circuit can exist by itself; it has several applications, such as a flashlight, a search light, and so forth.
>
> An example of a complicated circuit would be a radio receiver in which there are several elements with many connecting paths. Although a radio receiver seems complicated, this circuit can be analyzed using the techniques covered in this book. Our goal in this text is to learn various analytical techniques and computer software applications for describing the behavior of a circuit.
>
> Electric circuits are used in numerous electrical systems to accomplish different tasks. Our objective in this book is not the study of various uses and application of circuits. Rather our major concern is the analysis of the circuits. By the analysis of a circuit, we mean a study of the behavior of the circuit: How does it respond to a given input? How do the interconnected elements and devices in the circuit interact?
>
> We commence our study be defining some basic concepts. These concepts include charge, current, voltage, circuit elements, power, and energy. Before defining these concepts, we must first establish a system of units that we will use throughout the text.
>
> As electrical engineers, we deal with measurable quantities. Our measurement, however, must be communicated in a standard language that virtually all professionals can understand, irrespective of the country where the measurement is conducted. Such an international measurement language is the International System of units (SI), adapted by the General Conference on Weights and Measures in 1960. In this system, there are six principal units from which the units of all other physical quantities can be derived. Quantities and units are length (meter), mass (kilogram), time (seconds), electric current (ampere), thermodynamic temperature (Kelvin) and luminous intensity (candela).

Chapter resource 4.2

Writing Skills Reference	
Code	Comments
GT	Use requested title and/or sub-titles for added clarity
GF	Use 11 or 12 point font for professional appearance
GP	Use page break to prevent splitting bullet points or paragraphs between pages
PB	Use parallel construction of bullets (start all with a noun or all with a verb, but do not mix)
SS	Work on sentence structure–a sentence expresses a complete thought
SL	Replace long, run-on sentences with two or more shorter sentences
SR	Reduce number of sentences that start with the same word
PS	Improve paragraph structure (typically, a paragraph has a topic sentence and supporting sentences)
PL	Break long paragraphs into two or more shorter paragraphs for added clarity
DS	Double space between paragraphs and single space within paragraph
SV	Review principles of subject/verb agreement
PT	Review correct use of punctuation
AR	Review correct use of articles (a, an, the)
AF	Review correct use of "affect" and "effect"
IT	Review correct use of "it's" and "its"
TN	Review correct use of "than" and "then"
TR	Review correct use of "their" and "there"
TO	Review correct use of "to" and "too"
WH	Review correct use of "who" and "whom"
WD	Improve word selection for clarity and reader interest
SP	Run Spellchecker and make corrections

CHAPTER 5

Professional Presentation Skills

Information Transfer
Developing Professional Presentation Skills

"Effective presentation skills are key elements towards a successful engineering profession."

—Marius Marita, PE, Electrical Engineer and Advanced Engineer at FirstEnergy Corporation

PROFESSIONAL PRESENTATION SCENARIO

"My path down a 24-year career has been full of professional presentations that started with a single opportunity during my senior year in college. That single experience was not only career-changing but also life-changing. My speaking experiences have allowed me to progress in my career from a college student to project engineer, project manager, operations manager, product line manager, to a general manager running a division in a major company. Writing and presentation skills can be directly related to the opportunities I have been able to accept over my 24-year career."

(John J. Paserba, General Manager, Gas Circuit Breaker Divison, Mitsubishi Electric Power Products, Inc.)

Some of your most important opportunities to prepare for career success are found outside of your formal classes. This was true for John Paserba, and during our conversation with him at the end of this chapter, we will learn how this single event at Gannon University changed his life.

5.1 Introduction

The purpose of this chapter is to focus on the important nontechnical skills you need to develop and deliver professional presentations. Although written communication continues to expand with the use of portable electronic devices, it is likely you will have many opportunities to give presentations to audiences of peers, managers, clients, and community groups. In addition, you may be involved in Web-based conferences and seminars.

LEARNING OBJECTIVES

By using the information and exercises in this chapter you will be able to:

- Appreciate why the ability to speak before an audience is important to career success.
- Use the fundamental principles of public speaking for your presentations.
- Learn how to plan and deliver effective presentations for different types of audiences.
- Prepare and use effective visuals.
- Build confidence and deliver outstanding professional presentations.
- Participate effectively in team presentations.
- Minimize the fear of public speaking by gaining speaking experience.

Your ability to communicate ideas and project details in live presentations before an audience or in Web-based settings has a significant impact on your career success. Many technical professionals have excellent information, but if they cannot effectively deliver this to others during presentations and group discussions, they will fail to achieve desired results.

In addition to discussing the principles of public speaking, this chapter uses exercises to show you how to use engineering tools to enhance your presentation skills. This approach will save you time, reduce stress and fear, and help you deliver effective and professional presentations.

5.2 Public Speaking

Next to listening, the greatest amount of time in communicating is speaking. You may be reasonably comfortable in informal conversations with friends and peers. But, when you stand before a group of people and speak, you may feel some anxiety, as demonstrated in Figure 5.1.

Many engineers and other technical professionals are uncomfortable when involved in public speaking, at least until they have gained experience by giving several presentations. Some have said they fear speaking before a live audience more than death. This sounds rather severe, but indicates the need to build confidence by improving speaking skills.

As professional engineers, the authors of this book have delivered several thousand lectures and other presentations. It is easy for us to give presentations now, but the first few were difficult. Our engineering experience helped us develop speaking skills, and we will show you how you can use the same skills to become an effective speaker.

This practical approach, developed by engineers for engineers, should be easy for you to use. As engineering students, you can relate to this structured process to develop and deliver professional presentations. This will save you time in preparing visuals, and help you build public speaking confidence.

Figure 5.1
Fear of public speaking.

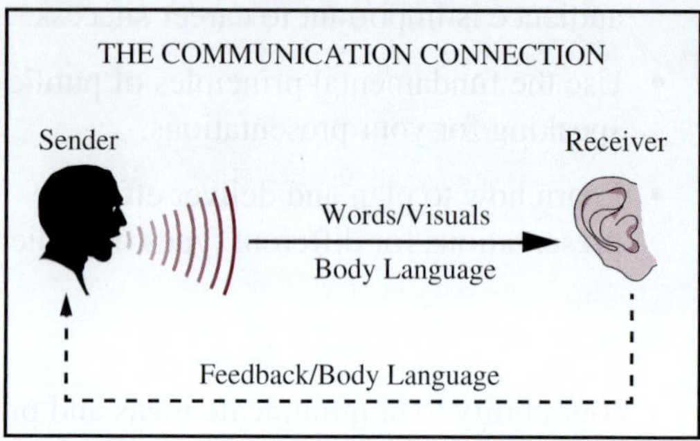

Figure 5.2
The communication loop.

> ***"A great engineering design or analysis may be rejected if it is hidden behind a poor presentation. Every project that an engineer works on will require the approval of another party. If that party does not understand the benefit that will be received from the associated cost, then the project will never proceed. As a result, the ability to give a good presentation is a critical skill for an engineer to possess."***
>
> (Paul Kruger, Electrical Engineer in an electric utility industry)

As illustrated in Figure 5.2, communication is more than just sending a message. The message needs to be delivered in an appropriate style so that receivers process information as they listen. Receivers may also give important feedback.

Effective speakers are good communicators. They understand and use the principles of public speaking. The secret to successful speaking techniques is to adapt principles to personal styles. Speaking style includes interaction with visuals and the audience, and the use of your voice to vary delivery levels and speed.

> The principles of public speaking are planning and preparation, practice, and professional presentation.

5.3 Planning and Preparation

Professional presentations require sufficient time for planning. An effective speaker will invest a minimum of twenty-five to thirty times the amount of time involved in actually delivering a presentation to develop, rehearse, and become familiar with information and visuals. The planning process should be started early and include sufficient time to develop and rehearse the presentation.

Start planning activities by identifying your audience. Because the audience has a great impact on your ability to deliver your message effectively, you should design your presentation to fit their needs and level of technical expertise. If necessary, ask questions to determine who will be in the audience and why they are interested in your topic.

5.3.1 Identifying Specific Audience Needs

To save preparation time and be more effective with audiences, you should think about why your information is important to various audiences and then determine the best way to communicate your message. Typical audiences in your future may include:

- Supervisors and peers
  - Are interested in technical information and many facts
  - Understand technical terms
  - Understand detailed sketches, graphs, and diagrams
  - Ask technical questions
  - Are comfortable with longer, detailed presentations
  - Relate to informal presentations and animation to demonstrate processes
- Senior management
  - Are less interested in technical information
  - May not understand technical terms
  - Can be overwhelmed with too much detail
  - Focus on economics and what technology can do for the bottom line
  - Usually ask nontechnical questions
  - Are extremely busy and have limited time to listen to presentations
  - Want brief, concise, and more formal presentations
  - Are not impressed with excessive animation in PowerPoint presentations
- Clients
  - Can have a combination of technical and nontechnical backgrounds
  - Want to know some technical detail
  - Are interested in how the project helps them and their bottom line
  - May ask both technical and nontechnical questions
  - Want brief but informative presentations
  - Are not impressed with excessive animation in PowerPoint presentations
- General public and other nontechnical audiences
  - Are less interested in technical information
  - May not understand technical terms and usually ask nontechnical questions
  - May be overwhelmed with too much detail
  - Want brief, concise presentations
  - May or may not be impressed with animated PowerPoint presentations

A classroom audience is similar to supervisors (instructor) and peers (students). Senior design project presentations may also include supervisors, peers, and clients (visiting industrial representatives). Students who attend professional society meetings are peers.

All presentations are important, and you should strive to deliver your message in the most appropriate way for each type of audience. However, presentations to senior management are the most critical and you should plan them very carefully. Senior managers expect you to be informative, clear, and concise and to end on time. Remember, your career future often depends on your ability to have a positive impact on senior management.

Start your preparation by making an outline of major points. Use a word processor to structure the presentation and arrange ideas for a logical flow of information. Continue by adding details to each section of the outline. Prepare introduction and summary sections after the main body of the presentation has been completed.

Select the type of presentation and visuals that will be most appropriate for the audience. Use major points of the outline to structure visuals. Then, use visuals to practice your presentation.

After preparing visuals, rehearse by talking aloud about information covered in each visual. Think more about overall concepts than specific words.

Time rehearsals, and adjust the amount of information to fit the amount of time allocated for your presentation. A general rule for time use during the presentation is to allocate 10% for an introduction, 80% for the main presentation, and 10% for the summary.

5.3.2 Preparing Effective Visuals

You can use a variety of visuals, depending on the audience and type of meeting. Options for visuals include black/white boards, flip charts, DVDs, and PowerPoint. Most visuals are prepared before the meeting and provide an outline for the discussion. Our discussion will focus on PowerPoint visuals, but you can apply many of the same ideas to other media as well.

Content In addition to knowing what to say, the next most important point is to support your comments by preparing visuals that are easy for the audience to see and understand. It is better to use a few, common words on a PowerPoint word visual instead of sentences, paragraphs, or long quotations. Your audience will remember more of what you say when they see key words and you provide additional details in your comments.

> Use the KISS principle to prevent barriers.

A general rule for all visuals is the KISS principle, "*K*eep *i*t *s*imple, *s*tudent." A common mistake is to clutter visuals with too much written information or detailed sketches and diagrams. Your audience can understand visuals more quickly when you limit the amount of information on each one. A good example is shown in Figure 5.3. This is much easier to read and understand than Figure 5.4. Unfortunately, too many visuals look somewhat like Figure 5.4 and are too busy for the audience to read and understand.

MULTI-MEDIA COMPUTERS

- ◆ Dynamic Visuals
- ◆ Convenient To Use
- ◆ Flexible For Changes
- ◆ Interesting For Audience
- ◆ Professional Image
- ◆ Portable For Travel

Figure 5.3
Effective word visual content.

MULTI-MEDIA COMPUTERS

- Dynamic visuals that can be prepared before the presentation and that will be a good guide for the speaker to use during the discussion and questions
- Convenient to use because they are easy to carry and easy to show to the audience so they can follow the words being said
- Flexible for changes in that you can make corrections or changes just before the presentation to be more current and appropriate for the audience
- Interesting for audience to see something while you are talking about the details of each topic as you proceed through the presentation
- Professional image that demonstrates that you know how to use PowerPoint and can deliver your message in a modern way for your audience to see
- Portable for travel because you can take these in your computer or you can just save the message on an USB memory stick, CD, or other memory system
- Expensive to own but this is usually worth the cost because it makes it much easier to prepare and deliver your message in a professional way
- Available everywhere and audiences are familiar with this type of presentation visual support so will be happy to see your visuals
- Reliable but some older systems for projection may not be compatible and require you to change the resolution of your computer to fill the screen
- Effective for presentations because the audience can read more details as you are talking about each topic
- Common to most speakers and most speakers know how to use this media
- Very useful tool to help you deliver many details about your presentation and you do not need to have a lot of visuals to tell your story-art equipment and also that you are up-to-date with all the
- Sign of professional speaker, especially when you put a great deal of information on each visual to show how much you know about your topic
- Demonstrates state-of-the wonderful electronic presentation systems
- Expected by most audiences because they have seen many PowerPoint presentations and always expect to be entertained in addition to hearing the story
- Can be distracting if a lot of animation is used to add emphasis and to direct your audience to each part of the presentation
- Needs a projector but these are usually very easy to get and most hotels and companies have several of these available at all times
- Provide excellent notes for speaker if you can see what is on the very busy visual and then use this to help you remember what to say and when to say it
- Can also be used as an eye test by your audience to help them determine if they need to get new glasses or if their present glasses are fine
- Although you can put a lot of information on one visual, you eventually run out of space and have to go to the next visual

Figure 5.4
Ineffective word visual content.

Format for Presentations to Live Audiences When preparing PowerPoint visuals to be projected on a screen in front of a live audience, use dark colors and simple designs for backgrounds. A long series of white backgrounds tend to cause eye fatigue. Also, when presentations are video-recorded, white backgrounds reduce camera shutter openings and images of speakers are often too dark. Dark backgrounds provide excellent contrast when you select white, yellow, and other light colors for words, graphs, and other figures.

As indicated in Figure 5.5, 44-point, white, Arial fonts are appropriate for titles. Most titles should be limited to one or two lines so they do not take up too much space. When you use all capital letters and add bold and shadow features, words are easier to read.

When using bullet points, we recommend 40-point, yellow, Arial fonts, with bold and shadow features for the first level of information. Second-level bullets can be 36-point white or other light color Arial fonts. Continue using bold and shadow features and capitalize the first letter of each word for all bullets. Avoid third-level bullets because they are small and difficult to see.

Minimize information overload by limiting bullet points to six lines of information under the title. If information does not fit in six lines, you can make a series of visuals with the same title.

Pictures and graphs are an excellent way to enhance visual presentations. Select pictures with limited details so your audience can quickly understand them. Graphs should be simple and uncluttered. For added clarity, use the power of 10 for large values on the *X* and *Y* axes. Select appropriate graph styles so the entire audience can easily understand the information.

Charts and graphs with properly selected colors are most effective. Subconscious minds tend to remember graphic relationships longer than columns of numbers. The large amount of information in Figure 5.6 is a barrier to the message. The graph in Figure 5.7 is easy to understand and is more effective.

You can use block diagrams, similar to the sample in Figure 5.8, to introduce more complicated information and data. These can be animated to demonstrate the flow of activities or processes. Block diagrams give your audience the big picture and prepare them for details. Then, you can talk about details of each block area with additional visuals after showing the overall process.

44 POINT, WHITE, BOLD, SHADOW FONT
ALL CAPITAL LETTERS-1 TO 2 LINES

- 40 Point, Yellow, Bold. Shadow Font
- Capitalize lst Letter Of Each Word
  - 36 Point, White, Bold Shadow Font
  - Capitalize lst Letter Of Each Word
- Dark Backgrounds
- Limit To Maximum Of 6 Lines

Figure 5.5
Effective word visual for projection on a screen.

RESULTS OF OIL CIRCUIT BREAKER TESTS AT CLOVERDALE SUBSTATION

123.456	124.456	132.222	133.432	150.002	108.322	98.566	108.322	96.666	111.222
144.332	99.003	100.110	98.752	99.332	100.000	102.224	103.333	102.444	108.662
123.456	124.456	132.222	133.432	150.002	108.322	98.566	108.322	96.666	111.222
133.466	133.486	122.226	132.422	148.001	146.522	133.333	132.222	131.112	130.220
123.456	124.456	132.222	133.432	150.002	108.322	98.566	108.322	96.666	111.222
144.332	99.003	100.110	98.752	99.332	100.000	102.224	103.333	102.444	108.662
133.466	133.486	122.226	132.422	148.001	146.522	133.333	132.222	131.112	130.220
123.456	124.456	132.222	133.432	150.002	108.322	98.566	108.322	96.666	111.222
144.332	99.003	100.110	98.752	99.332	100.000	102.224	103.333	102.444	108.662
123.456	124.456	132.222	133.432	150.002	108.322	98.566	108.322	96.666	111.222
144.332	99.003	100.110	98.752	99.332	100.000	102.224	103.333	102.444	108.662
123.456	124.456	132.222	133.432	150.002	108.322	98.566	108.322	96.666	111.222
133.466	133.486	122.226	132.422	148.001	146.522	133.333	132.222	131.112	130.220
123.456	124.456	132.222	133.432	150.002	108.322	98.566	108.322	96.666	111.222
144.332	99.003	100.110	98.752	99.332	100.000	102.224	103.333	102.444	108.662
133.466	133.486	122.226	132.422	148.001	146.522	133.333	132.222	131.112	130.220
123.456	124.456	132.222	133.432	150.002	108.322	98.566	108.322	96.666	111.222
144.332	99.003	100.110	98.752	99.332	100.000	102.224	103.333	102.444	108.662
133.466	133.486	122.226	132.422	148.001	146.522	133.333	132.222	131.112	130.220
123.456	124.456	132.222	133.432	150.002	108.322	98.566	108.322	96.666	111.222
144.332	99.003	100.110	98.752	99.332	100.000	102.224	103.333	102.444	108.662
133.466	133.486	122.226	132.422	148.001	146.522	133.333	132.222	131.112	130.220
123.456	124.456	132.222	133.432	150.002	108.322	98.566	108.322	96.666	111.222
144.332	99.003	100.110	98.752	99.332	100.000	102.224	103.333	102.444	108.662
123.456	124.456	132.222	133.432	150.002	108.322	98.566	108.322	96.666	111.222
144.332	99.003	100.110	98.752	99.332	100.000	102.224	103.333	102.444	108.662
123.456	124.456	132.222	133.432	150.002	108.322	98.566	108.322	96.666	111.222
133.466	133.486	122.226	132.422	148.001	146.522	133.333	132.222	131.112	130.220

Figure 5.6
Ineffective data visual.

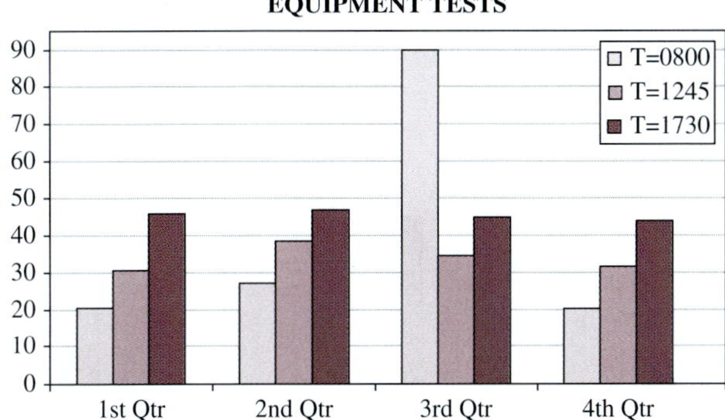

Figure 5.7
Effective data visual.

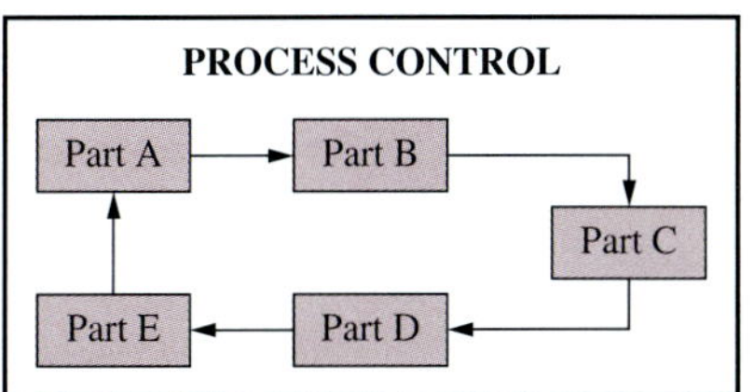

Figure 5.8
Sample block diagram.

Format for Web-Based Presentations Most of the principles relating to visual format for live audiences also apply to Web-based situations. This is certainly true for the amount of information and size of fonts.

However, because Web-based visuals are usually viewed on computer screens, you can use colors that may not be effective on a large screen in live audience presentations. For example, black on white word sides and graphs with white backgrounds can be effective on electronic screens. This does not result in eye fatigue because the screen is much smaller. You can also use as many as eight lines of bullets in Web-based visuals. Figure 5.9 shows examples of Web-based visuals that will be easy to read and understand.

Animation PowerPoint software includes a variety of visual animation and special effects. Some of these can be incorporated into presentations to add interest and emphasize major points. However, you should resist the temptation to make your presentation too entertaining. This is a frequent complaint from technical audiences because it tends to distract from the message.

When giving presentations to live audiences, you have the option of using a pointer or laser to interact with visuals on a screen. We discuss the pros and cons of pointers in the section on professional presentations.

A more professional approach to visual interaction is to develop visuals with overlays and progressive bullets. While animation should not be overdone, it offers an excellent way to emphasize major points of your presentation and keep audiences in step with your comments.

Progressive bullets help audiences follow your discussion point by point and prevent them from jumping ahead of your comments to additional bullets on the visual. Overlays of arrows, box outlines, shading, and other effects can be used to emphasize various parts of a visual.

Typically, progressive bullets and overlays are most effective with live audiences. It may be more difficult to control animation in Web-based presentations, especially during team presentations if only one team member controls transitions and animation.

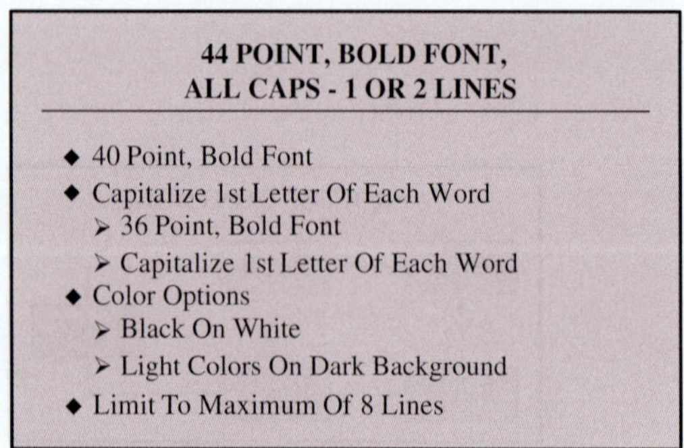

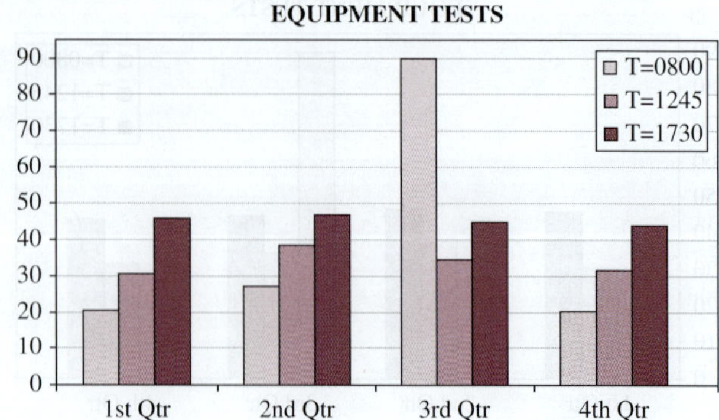

Figure 5.9
Appropriate Web-based visuals.

The number of visuals is a personal choice. If one visual is on the screen more than 30 to 60 seconds, it may become boring. On the other hand, the visual should be on the screen long enough for the audience to fully understand its content. When visuals are simple, the audience needs less time to comprehend the message.

General Comments Relating to Visual Preparation To make your presentations more interesting, include pictures and graphs along with word visuals. Digital cameras are a convenient way to take pictures of events and projects. "A picture is worth a thousand words" applies to presentations. Pictures add variety and help your audience to be more attentive during your presentation.

Preparing computer visuals can be time consuming. A reasonable amount of animation can be included if development of the program starts early. So, start your preparation of visuals early and consider how you can make them more effective to help you deliver your message.

Because computer visuals offer a multitude of options, it is easy for the audience to focus on an "entertaining" delivery and miss the content of the message. Pictures should have a direct relationship to the message and add to the presentation without becoming a distraction. It is more important to keep visuals easy to understand than it is to entertain with special effects.

5.3.3 Practice

You can develop a more professional presentation by investing a significant amount of time to rehearse your presentation. One proven technique is to practice what to say with each visual. You can build speaking confidence by scheduling time for several practice sessions.

If you rehearse aloud you can practice projecting your voice and developing a smooth delivery. Effective speaking style is developed by being confident, by using positive body language such as looking at the audience and using natural gestures. Improve your delivery by including variations in your voice level and speed, and by eliminating time fillers such as uh, you know, and next between sentences. Use your practice sessions to develop a speaking style that will be interesting and that will encourage the audience to listen and enjoy your presentation.

> Practice delivering general ideas instead of memorized words.

Practice talking about the content of each visual without memorizing specific words. If you memorize a presentation, it is easy to forget what to say if you are nervous when you stand in front of an audience. Instead, use the information on visuals to help you remember what to discuss, and then just talk about overall concepts of each topic.

5.3.4 Presentation—Delivering the Message

Now, let's discuss the major part of any presentation—delivery.

Presentations can either build a positive image for you or result in negative impacts on your career. It all depends on how you deliver your message. You can demonstrate your knowledge and enhance your image by making your presentations more professional. And the good news is that you can improve your presentations skills now by taking advantage of many opportunities to practice while you are in school.

Presentations to Live Audiences

Introduction

When starting a presentation, your introduction should establish a connection with the audience. You can do this by standing near the audience, establishing good eye contact, and giving appropriate comments in your introduction. Audiences are usually more interested in what you have to say and how it relates to them than they are in you. So, use introductory comments to explain why your message is important to them. The introduction should take about 10% of your total presentation time.

Eliminating Barriers

The effectiveness of any presentation is significantly improved by eliminating barriers between you and the audience. Barriers include a variety of ineffective presentation visuals and poor speaking styles that cause the audience to lose interest and think about something else.

Eliminate barriers between you and the audience.

Many barriers can be eliminated by good planning and by preparing visuals that are easy to read and understand. When you use visuals for notes, you can focus on the audience and increase eye contact and voice projection. Properly designed visuals also increase audience interest and become a bridge to the presentation content and not a barrier.

Dress appropriately for the audience. If the audience focuses more on how you look than what you say, it can be a barrier and distract from your message. You can dress more casually for peer presentations, but should use business attire for senior management and clients. It is usually better to dress casually for most community and nontechnical audiences.

While it is good to interact with visuals to help the audience focus on what is being discussed, some interaction can also become a barrier. For example, if you use a pointer and face the screen with your back to the audience as you point to the visual, this becomes a major barrier. Your eye contact and audience focus will disappear and your voice will be projected to the screen. This can be avoided by placing the pointer in your hand next to the screen, pointing to a specific area on the screen, and then looking at the audience to discuss the information.

Similar barriers are associated with using a laser pointer. If you are nervous, your shaking hand will be magnified by the laser beam on the screen. In addition, you tend to face the screen and lose audience contact when pointing to the visual with a laser.

Other barriers are caused by poor speaking style or nervous movements that are often the result of limited experience. Frequently, speakers use time fillers as they are thinking about what to say. Repeating titles and bullets on the visual to the audience becomes a barrier.

PUBLIC SPEAKING BARRIERS

- Improper dress (overdressed or too informal)
- Busy visuals with too much information, poor contrast or small fonts
- Negative attitude and body language
- Obvious nervous movements
- Rocking from one foot to the other or continually waving hands
- Being stiff like a statue and not using natural gestures
- Hiding behind the lectern or other physical barriers
- Low energy level and lack of enthusiasm
- Chewing gum
- Lack of eye contact with the entire audience
- Inappropriate level of information for specific audience
- Speaking too fast or too slow
- Lack of voice projection to entire audience
- Repeating information from visuals, word for word, to audience
- Saying "thinking time" fillers such as uh, you know, now, and next
- Reading presentation to audience from note cards or script

Figure 5.10
Typical presentation barriers.

By recognizing these barriers, you can remove them by better preparation, additional rehearsal time, practicing in front of a mirror, and speaking experience. Fortunately, you can gain this experience now, as a student, by taking advantage of speaking opportunities in classes, laboratories, and professional organizations.

Typical barriers to delivering a message are summarized in Figure 5.10. If the audience remembers any of these barriers instead of your message, you have failed to communicate effectively. Review this list before giving your next presentation, and work to eliminate barriers you may have used in previous presentations.

Using Visuals for Notes

Unless your topic is a legal document or you are representing your company before a congressional committee or other high-level audience, avoid reading to the audience. When you hide behind a lectern and read your paper word for word, your presentation will be boring and ineffective, as demonstrated in Figure 5.11. If you observe the

Figure 5.11
Boring and unprofessional delivery.

Figure 5.12
Professional delivery.

audience while reading, you will find that eyes are closing, heads are nodding, and eventually you may even hear snoring sounds.

But you may say, "There is no way I can face an audience without using notes or reading my presentation." Well, if that is how you feel, we have a very practical solution for you. You can actually use notes and be quite professional. The secret is to use your visuals for notes, as the speaker in Figure 5.12 is demonstrating.

Develop visuals to include major points that you want to discuss and practice talking about the concepts outlined on your visuals. Then, you will not need written notes, and you can build confidence by using information on your visuals to remember what to say.

> ***"The presentation skills developed during my senior year at Cleveland State University helped me put together effective presentations throughout my 5-year career in power systems engineering. I am thankful for the guidance and support of my instructors who emphasized the value of good communication and showed me how to improve my public speaking skills."***
>
> (Marius Marita, PE, Electrical Engineer and Advanced Engineer at FirstEnergy Corporation)

In addition, visuals help your audience understand and remember what you say. Without visuals, retention of information is limited. Using good visuals increases retention to

TYPICAL INFORMATION RETENTION	
Presentation without visuals	10%
Presentation with good visuals	40%
Presentation with good visuals and handout materials	100%

Figure 5.13
Typical retention rates for presentations.

as much as 40%. When you distribute printed summaries of your major ideas, your audience has the opportunity to increase their retention up to 100%, as shown in Figure 5.13.

Handout materials are usually distributed after the presentation. However, if you want audience participation, such as in workshops, give worksheets to the audience at the beginning and use them as a reference during the presentation.

The use of properly selected and designed visual aids will multiply the effectiveness of your presentation. Even more importantly, when you use visuals as a guide, you will build confidence and minimize feelings of anxiety.

Presentation Summary

The last 10% of your allocated time should be used to summarize your presentation. This is an excellent way to emphasize the most important points of your subject. Your audience will remember information in your summary longer because they hear it near the end of the session.

The summary is another opportunity to show how your topic applies to the audience. Your summary is also a good way to initiate questions. This brings us to the last part of most presentations—a question-and-answer session.

Responding to Questions

An important part of most presentations is the opportunity to answer questions from the audience. So, how can you handle questions effectively and avoid being intimidated? One approach is to tell the audience during your introduction when you would like to have questions.

Questions can be asked during the presentation or at the end. Answering questions during the presentation can be time consuming and, in some cases, distracting. However, if the presentation is a workshop and you want audience participation, you can encourage the audience to ask questions during the workshop.

A second way to answer questions effectively is to think about potential questions as you prepare for the presentation and develop answers to them before the event. Then, it will be much easier for you to handle these questions if they are asked by the audience. Additional tips on handling questions are shown in Figure 5.14.

Team Presentations

Because you will probably be a member of a team when involved in large and complex projects, you may also be part of team presentations. All of the principles discussed to this point apply to team presentations as well. But, there are additional points you need to keep in mind when giving team presentations.

RESPONDING TO QUESTIONS

- Repeat or paraphrase the question to be sure you and the audience heard it correctly
- If you know the answer, give a concise reply
- If you need time to think about an answer, you can say, "That is a good question -let me think about it for a moment"
- If you don't know the answer, admit it, and then ask the person to see you after the presentation and discuss how you can provide the answer at a later date
- If you receive a question that disagrees with your comments, you can answer in a professional way by saying there can be more than one opinion and you expressed your opinion in the presentation

Figure 5.14
Tips for responding to questions.

In addition to delivering your part of a team presentation, if you are the first speaker, you have the responsibility to introduce the team and project. A good introduction will set the tone of the team presentation and develop audience interest and attention.

Key activities for the first speaker include

- Avoiding speaking until first looking at the audience to establish an audience connection.
- Introducing the project and team members, and identifying who will be covering the various parts of the presentation.
- Presenting appropriate details using good eye contact and voice projection.
- Completing the presentation with a brief summary of major points.
- Making an effective transition to the next speaker by giving his or her name and topic, and perhaps using a transition visual.

The transition will be most effective if the next speaker uses transition statements to bridge from the previous topic to the next. Key activities for team members who follow the first speaker are

- Starting with a brief introduction of his or her part of the presentation.
- Giving appropriate details with good eye contact and voice projection.
- Completing the presentation with a brief summary of major points.
- Making an effective transition to the next speaker.

The last speaker has an important task in addition to delivering his or her part of the team presentation. The final speaker should summarize the major points of the total team project to emphasize the most important information, and set the stage for questions from the audience.

The suggestions on how to answer questions as an individual speaker also apply to questions for teams. One additional suggestion for team members is to decide who will answer questions and how they can work together so they do not talk over each other in giving answers. In all aspects, individual parts of team presentations should be coordinated to result in one, overall presentation.

A team presentation should be a single, coordinated discussion and not a series of individual presentations.

Mechanical Problems Even with good planning, audience analysis, and preparation of visuals, mechanical or technical problems may surface during your presentation. For example, the room arrangement may be inappropriate for the audience to see the screen or lights may not be easily controlled.

Arrive early to check out equipment.

Additional problems occur when microphones do not operate properly or when projector bulbs burn out during presentations. If you are using an on-site computer instead of your own, your version of PowerPoint may not be compatible and animations may not work properly.

To prevent many of these problems, arrive early and check out as many systems as possible before your presentation. As a speaker, it is your responsibility to arrange for proper equipment to fit your needs and to minimize audience distractions. Request equipment before the meeting and then arrive early enough to practice using it before your presentation begins.

Web-Based Presentations In many respects, Web-based presentations are similar to live presentations. In both cases, the purpose is to transfer information from speakers to the audience, and to do this in a way that the members of the audience can understand and apply the message to their needs.

Many speakers fail to communicate effectively because they let barriers interfere with the delivery of their message. This is especially important for Web-based presentations because communication tools are usually limited to visuals and the speaker's voice. So, it is important to maximize the use of these tools and minimize barriers in the process.

Typical barriers for Web-based visuals are

- Too much information on a single visual.
- Each visual on the screen too long, resulting in listener distraction.
- Poor contrast between information (words, diagrams, graphs, charts, etc.) and visual background caused by poor color selection or busy backgrounds.
- Small font size that is difficult to read.
- Lack of coordinated interaction between the visual and spoken comments.

Typical barriers associated with the speaker's voice are

- Using low voice levels and lack of projection.
- Having poor-quality microphones or being too far from the microphone.
- Using time fillers.

- Having a lack of enthusiasm and little change in voice levels to emphasize major points.
- Using a delivery that is choppy with long pauses between visuals or major thoughts.
- Reading the material or repeating information on the visual to the audience.

For best results, design visuals to eliminate barriers. Increase rehearsal time and practice effective interaction with visuals. Avoid memorizing specific words, but practice discussing concepts relating to each visual. To deliver professional presentations, do not read from a script or repeat titles or words from the visual for the audience.

If possible during Web-based presentations, arrange to personally advance visuals and then you can use progressive bullets, overlays, and other animation to correlate interaction of visual details with your comments. This is usually more professional than using the mouse or pointer, which can become a distraction.

Rehearse aloud to develop voice projection and a style that includes various speeds and levels. Use your voice to emphasize major points and to be more interesting and professional. Eliminate time fillers and develop a smooth delivery. Although you may not be visible in online presentations, you can use your voice to demonstrate confidence and your knowledge of the subject.

Some Additional Thoughts A discussion of major points on how to deliver a professional presentation is a good way to finish this part of the chapter. So, here are some of the most important tips for success.

To relax, take a few deep breaths just before moving to the front of the audience. Start the presentation by standing near the audience and not behind the lectern.

Focus on *ideas* and not specific words. Deliver information without written notes and use visuals for notes. Locate your computer screen or monitor where you can easily see it and use this to view your visuals and not look at the screen behind you. Then, you can look more at the audience and less at the screen.

Prepare your presentation for the benefit of your audience and monitor their response during the presentation. When you see signals the audience is restless or bored, make changes in your delivery style.

Proper use of visuals is one of the best methods to keep your audience interested. Use a mix of words, graphs, and pictures, and include an appropriate amount of information.

Voice levels and speed can be changed to add variety to your presentation. Use style, body language, and movement to demonstrate enthusiasm.

Some forms of humor can be effective, but humor can also be risky. Keep all humorous stories short and relate them to the presentation. Never use humor that could offend someone in the audience. When in doubt, leave it out.

Finish with a summary of the most important points of the presentation. This helps your audience remember information and may encourage them to ask questions to expand the value of your presentation.

Be flexible and use only the amount of time allocated for your presentation. You may be notified of a change in your time allocation just before you are introduced, so adjust your presentation to end on time. It is always better to end early than to go over the time limit.

5.4 Value of Public Speaking Experience

When it comes to public speaking, there is really no substitute for experience. Look for opportunities to speak to a variety of audiences to gain experience and to build confidence. Use friendly environments, such as classes, laboratories, and professional organizations to practice speaking to audiences.

Professional organizations offer excellent opportunities for you to practice presentation skills. As an active member, you can volunteer to introduce a speaker or give informal presentations before an audience of your peers. You can also gain additional skills and self-confidence by leading projects or being an officer in these organizations. Other organizations, such as Toastmasters International (www.toastmasters.org), are located in most areas and offer excellent opportunities to gain speaking experience and confidence.

> ***"I used to be an introvert and dreaded giving presentations. However, I knew that the ability to present ideas was an integral part of a successful engineering career. To overcome my fear I decided to participate as an officer in SWE (Society of Women Engineers). By regularly engaging in outreach events where I delivered engineering related presentations to small groups of high school students, I gradually became more confident speaking before a group.***
>
> ***When our local SWE chapter hosted a national outreach event, I was able to feel comfortable delivering the opening remarks in front of a crowd of over one hundred fifty students, parents and professionals.***
>
> ***Now, making presentations to corporate leaders is a regular part of my role as a Technical Advisor/Prior Art Search professional. Therefore, the public speaking experience gained through my active participation in a student organization has proved invaluable."***
>
> **(Maria Marez Baker, BEE, Technical Advisor, Electrical Arts, Frisina, LLC)**

You are encouraged to look for opportunities to speak to a variety of audiences to gain experience and to build confidence. Use friendly environments, such as your classes, laboratories, and professional organizations to practice speaking in front of an audience.

A CONVERSATION WITH JOHN PASERBA

John, I know you have given hundreds of professional presentations. How did you develop public speaking skills?

Answer

"I was a senior in college and our local IEEE (Institute of Electrical and Electronics Engineers) Student Branch was sponsoring a technical paper contest at my university supported by the local IEEE section. The contest consisted of a written paper and an oral presentation. As the date of the contest neared, the 'volunteers' for entering this contest were few. One of my professors encouraged me to enter the contest. I resisted at first but later, with proper academic encouragement (a.k.a. twisting of my arm), I agreed to enter the local contest.

I wrote a paper based on my senior design project and created a presentation. Back in those days, we used 'flip-charts' drawn by hand, as access to any computer-generated graphics programs, such as PowerPoint, was still off in the future, especially for college students. Four of us entered the contest and there were cash prizes for 1st, 2nd, and 3rd place.

I recall the evening of the contest very well. At the time, our university had a weekly 'Senior Seminar' that every graduating senior was required to attend. Normally the seminar was during the day, but, this particular week, we were all assigned to attend the evening IEEE dinner and presentation contest, so, the room was packed for the event.

The four of us presented our respective papers, and to my surprise I not only enjoyed the experience, but, I won first prize with the top cash award—a welcome addition to my sparse college income. Because I was the winner of a local contest, I was automatically invited to the IEEE Region Paper Contest, where a region consisted of several states and nearly seventy-five eligible schools with IEEE Student Branches. I updated my paper and my presentation based on the local contest experience, and traveled several hundred miles to participate in the Regional contest.

While I did not place in the top three of the Region Paper Contest, I did enjoy the experience so much that I decided when I graduated I would only take a job at a company that supported participation in a professional society (in my case, IEEE) and supported industry publications and presentations. This decision guided me in my graduate work and with the companies I interviewed with as I was nearing graduation.

When I graduated with a master's degree, I only considered taking employment with a company that offered this opportunity to publish. I was fortunate that I found such a company with which to begin my career. Later, when I moved to my second company, I pursued the same attributes of support of industry publications and presentations."

In this age of electronic communication, do you think it is important for engineers to learn how to give oral presentations in their working careers?

Answer

"There is no question in my experience that oral presentations are more important today than ever before. With e-mail, phone, text, PowerPoint, and many other media competing for people's attention, an opportunity for oral presentations can make the difference in getting your points, proposals, and ideas across to your colleagues and upper management. The key is to be brief, focused, and clear on your points. Flexibility is always critical. Multiple times

I have walked into upper management presentations and as I enter the room, I am asked 'can you make your points in one hour?' when the meeting was scheduled for two hours or more. Preparing short, crisp, concise, and on-point presentations is critical to a successful career."

To emphasize the value of public speaking skills, what are typical examples of your presentations in your work and in your professional activities?

Answer

"In twenty-four years, I have published and presented over sixty national and international technical papers in journals, conferences, and in magazines, including chapters in five *Engineering Handbook* editions. I have attended conferences and workshops and have made technical presentations around the world including in many cities in the USA and Canada as well as in Brazil, Mexico, Singapore, China, Japan, Belgium, Germany, Saudi Arabia, France, and Indonesia.

I have been invited to teach seminars and courses worldwide including the University of Pittsburgh, Pennsylvania State University, University of Wisconsin, Waseda University, Japan, Institute of Technology at Bandung, and others. Furthermore, I have been invited to speak at nearly forty colleges worldwide on engineering professionalism and career management. These presentations and others have been made to audiences from a dozen to thousands.

As a Project Manager, I had to make presentations to customers and to advisory boards on proposals for new business and for final reports on my own or my group's technical work. Later, as a Department Manager and as a Division General Manager I routinely make presentations to upper management to go over business plans (consisting of sales plans, staffing plans, operation budgets, and capital investments), and then to my employees for execution of those plans.

One of my presentations that had a high-impact occurred recently after an annual review of my business, which was presented to nearly three hundred professional, clerical, and factory attendees of my Division (I had been a Division General Manager for only 3 weeks at that time). After the 90-minute presentation, several of our factory technicians approached me and said 'I really comprehend our business and our customers and how it all comes together and the importance of my role—thank you for making it so understandable.' I'll never forget the impact of that moment—as it further emphasized the importance of professional presentations skills."

What advice do you have for students relating to professional presentation skills?

Answer

"The best advice I can give to students related to professional presentation skills aligns with my experience as an undergraduate and graduate student and a young engineer in my company when I first started: that is to take every opportunity you can to get professional speaking experience. Enter your university's paper or speaking contests and orally present your senior design project or other undergraduate projects. At your first job, if you are ever in small groups such as a seminar or workshop, volunteer to present your group's work. Your communication skills, both writing and speaking, will take you as far as your willingness to continually improve in this area."

(John J. Paserba, General Manager, Gas Circuit Breaker Divison, Mitsubishi Electric Power Products, Inc.)

5.5 Conclusion

Technical expertise is the foundation for your career. But, as John has clearly stated in our discussion with him, technical skills are only part of what you need to be a successful engineer. Success is achieved by a balance of technical and non-technical skills. And, communication is your most important nontechnical skill.

Remember, the purpose of making a presentation is to deliver appropriate information to an audience with a delivery process that makes it easy for them to understand and use the information. Effective and professional speakers chart their course early, analyze their audience, and prepare materials in time to permit adequate rehearsal before the presentation.

As noted in the beginning of this chapter, our objective is to encourage you to improve your speaking skills and to show you how you can do that. Your ability to communicate with technical and nontechnical people will be a key ingredient to help you reach higher levels of success and to make a positive impact on society. Use the principles of this chapter to develop communication skills now, and you will be better prepared to start your working career.

END OF CHAPTER REVIEW QUIZ

Select the most appropriate answer to the following statements.

1. The most important reason for you to improve your public speaking skills is
   - **a.** To receive a better grade when assigned to give a presentation in class.
   - **b.** To avoid being nervous when speaking before a large audience.
   - **c.** To prepare to run for a public office such as a school board member.
   - **d.** To deliver your ideas to others so they can understand and use them.

2. The best way to prepare to give an effective presentation before an audience is to
   - **a.** Spend most of the night before the presentation to practice.
   - **b.** Invest appropriate time to know your subject, make good visuals, and rehearse.
   - **c.** Spend time to make a lot of animated visuals so the audience will enjoy your presentation.
   - **d.** Memorize your presentation and prepare note cards to help you remember it.

3. The purpose of an introduction to a presentation is to
   - **a.** Outline what you are going to discuss.
   - **b.** Explain how your topic relates to the audience.
   - **c.** Start your presentation on a positive note so the audience will be interested.
   - **d.** Accomplish all of the above.

4. It is important to eliminate barriers between you and the audience because barriers
 a. Distract you and make it more difficult for you to remember what to say.
 b. Make it harder to recognize audience feedback.
 c. Distract the audience and cause them to lose interest in your presentation.
 d. Make the audience think you are a great speaker.

5. During your presentation, the best way to remember what to say is to
 a. Use visuals for notes.
 b. Read the presentation word for word so you don't leave out any information.
 c. Use note cards for guidance and read details from them to the audience.
 d. Look at the screen and repeat information from the visual to the audience.

6. You can reduce nervousness by
 a. Taking medicine to be more relaxed.
 b. Hiding behind a lectern so the audience can't see you.
 c. Rehearsing and using good visuals to build confidence in what to say.
 d. None of the above.

7. The most important value of summarizing your presentation at the end is
 a. To give you a way to end the presentation.
 b. To emphasize the most important points of the presentation.
 c. To use time so you don't have to prepare as much material.
 d. To discourage the audience from asking questions.

8. When answering questions the most important thing to do is to
 a. Repeat the question and answer in a concise way.
 b. Argue with someone who does not agree so you can defend your ideas.
 c. Write answers to potential questions and read the answers when the questions are asked.
 d. Answer the question by asking the audience what they think.

9. The best way to minimize mechanical problems is to
 a. Take your own audio visual equipment with you to the meeting.
 b. Discuss equipment needs several days before the event with the facility manager.
 c. Reserve equipment, arrive early, and test audio visual equipment and lighting controls.
 d. Have a someone adjust equipment during your presentation, if needed.

10. It is important to improve your presentation skills because
 a. This is a sign of a professional and can open career opportunities.
 b. You will save preparation time and reduce stress when speaking.
 c. You can make positive contributions to team project presentations.
 d. It includes all of the above.

EXERCISES TO DEVELOP AND ENHANCE YOUR SKILL SET

Exercise 5.1 Your instructor may select one of the following options for this exercise: 1) assign you to give a five-minute presentation to summarize a previous class lecture in class; 2) assign you to give a five-minute presentation to summarize a class lecture to one of your study groups. These assignments will include the use of evaluation reports.

Exercise 5.2 Change the busy and ineffective PowerPoint visual in Chapter resource 5.1 to one PowerPoint visual using bullets. Print this visual and give to your instructor on the date assigned.

Exercise 5.3 Prepare a PowerPoint presentation for the laboratory report in Chapter resource 5.2. Provide an electronic file of your PowerPoint presentation to your instructor, if assigned.

END OF CHAPTER RESOURCES

Chapter resource 5.1

Ineffective PowerPoint Visual

MULTI-MEDIA COMPUTERS

Multi-media computers offer many new tools to make visuals for presentations. These flexible tools provide opportunities for you to make presentations very dynamic. Softwere is user friendly, easy to learn, and convenient to use, and can be programmed with a variety of transitions and animation techniques. By using these format tools, you can make your presentations much more interesting for the audience. This helps you make a positive impact on your audience and results in a very professional image for you as a speaker.

Chapter resource 5.2

Laboratory Experiment to Demonstrate Ohm's Law

Introduction: This experiment was designed to prove Ohm's law by using a simple electrical circuit to measure voltage across resistors and current through resistors. The validity of Ohm's law formulas can be confirmed by comparing measured results to calculated values. Two resistors of different values were used to observe and record results of voltage and current.

Purpose: The purpose was to demonstrate that when resistance increases in a series circuit, the current decreases and equals the amount determined by Ohm's law that $I = V/R$, where I is current, V is voltage, and R is resistance.

Apparatus: A simple circuit was constructed by placing each resistor, one at a time, in series with a DC power supply. Resistor values were 1 kΩ and 3.3 kΩ. A voltmeter and an ammeter were connected at the proper locations in the circuit and used to measure voltage across the resistor and current through the resistor.

Procedure: The power supply was adjusted to create voltages across the resistor of 2, 4, 6, 8, and 10 V. At each voltage level, the current was measured. Results are tabulated in Table 1 for the 1-kΩ resistor and in Table 2 for the 3.3-kΩ resistor.

Calculated values for the five (5) voltage levels for the 1-kΩ resistor are shown in Table 1 and for the 3.3-kΩ resistor in Table 2. The difference (measured current less calculated current) is shown in the third column in both tables.

Results: As shown in Tables 1 and 2 , the differences between experiment results and calculated results using Ohm's law are small, proving that Ohm's law is correct. Differences can be attributed to small voltage drops within resistor wires.

MEASURED RESULTS			
VOLTAGE VOLTS	CURRENT (mA)		
	Measured	Calculated	Difference
0	0.00	0.00	0.00
2	2.00	2.04	−0.04
4	4.00	4.08	−0.08
6	5.98	6.12	−0.14
8	8.00	8.16	−0.16
10	9.90	10.20	−0.30

Table 1
Values for a 1-kΩ resistor.

MEASURED RESULTS			
VOLTAGE VOLTS	CURRENT (mA)		
	Measured	Calculated	Difference
0	0.00	0.00	0.00
2	0.62	0.62	0.00
4	1.22	1.23	−0.01
6	1.84	1.85	−0.01
8	2.45	2.47	−0.02
10	3.07	3.09	−0.02

Table 2
Values for a 3.3-kΩ resistor.

Conclusions: Experimental data confirmed that Ohm's law of $I = V/R$ is correct. The relationship of voltage and current is linear and current decreases when resistance increases.

CHAPTER

Information Capturing Skills

Your Personal Inventory

"My recommendation to students is to learn to document their activities and work. Students should realize that as practicing engineers, they will be held to much higher levels of accountability than 98% of the rest of the population."

—Elmer Paine, PE, Electrical Engineer

INFORMATION CAPTURING SCENARIO

"The last several years have resulted in much job instability in the engineering profession. To prepare for potential career changes, wise engineers need to 'keep their resume up to date.' So, how can you do that? What techniques can you use to keep a record of your professional experience and past job related activities? How can you successfully react to career detours? How is it possible to take advantage of unexpected opportunities?"

(Elmer Paine, PE, Electrical Engineer, and Sr. Electrical Engineer at Affiliated Engineers)

In this chapter, we answer these questions posed by a professional engineer, Elmer Paine. Then, we learn how he records and uses information in his engineering career in our discussion with him at the end of the chapter.

6.1 Introduction

Unlike past generations in which engineers worked for or owned a single company during their entire career, engineers today face a different and very dynamic future. Because you will probably work for more than one company, your career will include many options and outside forces that will play a significant role in determining your opportunities for success.

This raises some interesting questions. How do you prepare to make important decisions for career paths? In addition to your formal education and work experiences, what other resources do you need to move on if your job is terminated or when you

LEARNING OBJECTIVES

By using the information and exercises in this chapter you will be able to:

- Develop and use a personal marketing plan.
- Understand why documentation of experiences and skills is important.
- Develop and use an information capturing system (ICS).
- Use resources available from your University Career Center.
- Use your ICS to build superior resumes.
- Apply information from your ICS to secure future jobs.
- Use captured information to achieve success in a dynamic world.

want to change directions in your career? How can you be successful in your career and life?

We answer these questions by first discussing how companies use marketing plans to compete in the marketplace. Then, we relate these key marketing steps to a personal marketing plan. This will provide a foundation for you to learn how to capture information and then use it as a resource in your job searches and future job changes during your career.

6.2 Marketing Plans

We start with a focus on how you get a job. Some of your work opportunities may include coop, intern, or part-time jobs while you are a student. After graduation, you will be looking for full-time work. Your search for work will be more successful if you show prospective employers how your education and other experiences can be a valuable resource for them.

> Successful engineers develop a personal marketing plan.

When you seek employment, in many respects you are like a consultant and your potential employers are your clients. Consultants work for clients who can use their services, and clients compensate consultants for what they deliver to help clients improve their bottom-line profits.

Consultants use marketing principles to find customers and discuss their resources with the goal of obtaining a contract for work. You can use some of the same marketing principles when you are looking for employment. Let's see how your personal marketing plan compares to marketing activities of a consulting firm.

Consultants introduce their services to potential clients by developing a marketing plan. This plan includes identifying market opportunities, building an inventory

MARKETING 101	
CONSULTANT MARKETING PLAN	**PERSONAL MARKETING PLAN**
Skills and type of service	Personal skills
Customer research	Employer research
Inventory of resources	Portfolio
Description of services	Resume
Advertisement	Resume cover letter
Sales calls	Job interviews
Service delivery	Engineering contributions
Service support	Life-long learning

Figure 6.1
Marketing plans comparison.

of resources, creating a description of their services, advertising their skills, and using sales calls and other promotional activities to encourage clients to hire them. Once hired, consultants complete the process by delivering their services and providing long-term support to encourage future work opportunities.

You can use a personal marketing plan to find career opportunities. As shown in Figure 6.1, your personal marketing plan is like a marketing plan for a product or service. You can use marketing principles to help you obtain coop and intern positions and secure jobs after you graduate. So, let's see how you can use these principles to develop your personal marketing plan and sell your ability to be an important resource to a potential employer.

6.2.1 Personal Skills

For several years, you will be planning and preparing for your career by learning how to learn and by developing technical and nontechnical skills in classes, laboratories, projects, and professional organizations. Then, just like a consultant offering a service to a client, you can offer your personal skills to a potential employer. Your success in finding the job of your choice will depend on your ability to successfully discuss and promote your skills.

6.2.2 Employer Research

In the same way a consultant finds clients, you can identify potential employers by researching companies and by building personal networks. Use resources available at the university to start your research. Initially, you can focus on acquiring general information about companies who have products or services that interest you and who may be candidates for your future job search.

The information you obtain in your early research activities can be added to your personal marketing plan. Later, you can obtain more specific information about potential employers in these same industries. This information will save you time when preparing to seek employment.

6.2.3 Portfolio

Consultants develop an inventory of skills and resources much like companies that use warehouses to store products. Although consulting firms may not use a physical warehouse, they accumulate and "store" their knowledge, skills, and tools in anticipation of customer use.

In a similar way, you need an inventory of your skills and knowledge to prepare resumes and describe your unique qualifications to potential employers. This is called a portfolio and is a useful resource when seeking employment. When you start developing your portfolio early, you will have a large inventory of information.

A personal portfolio also gives direction to help you develop new skills to make your job search more successful. The portfolio structure indicates areas that need to be strengthened, and this offers you the opportunity to improve skills and gain experiences to build superior resumes. And that brings us to the next step in your personal marketing plan—your resume.

6.2.4 Resume

Unless you have work experience, you probably have not given much thought to a resume. Resumes are used to apply for coop and intern positions before graduation, and to help you obtain job interviews when starting your working career after graduation.

Your resume is very much like the description of skills and resources that consultants use to discuss what they offer to a potential client. It is a summary of what you offer to an employer and includes evidence of your knowledge and skills.

6.2.5 Resume Cover Letter

Consulting firms advertise their services to encourage potential customers to contact them for more information. In a similar way, the primary purpose of your cover letter is to advertise your resume and encourage potential employers to read it. So, the information in your cover letter should be specific to the company and position, and should demonstrate how you are uniquely qualified for that position.

6.2.6 Job Interviews

When you have a good resume, you will increase your chances for a job interview. The interview is much like making a sales call on a potential client. Good sales people are well acquainted with their products and they also know what their customers need. They prepare for sales calls and follow a plan to guide the discussion so it leads to a successful sale.

You can be a successful "sales person" when you plan for your interview by knowing what you can offer and how that fits the needs of a prospective employer. In this chapter, we discuss how you can start now to record information that will help structure your interviews. More details can be added as you gain additional experiences and acquire skills that will be important to discuss in your interviews.

6.2.7 Engineering Contributions

Consultants are well aware that future opportunities are based on the delivery of quality services. Some of the answers to questions raised at the start of this chapter include your delivery of professional work after joining a corporation or starting your own company. Outstanding work is important to increase security in job positions or to be prepared for job changes caused by your employer or career changes you may initiate.

6.2.8 Lifelong Learning

Consultants continue communicating with clients after they complete their work to make sure clients are satisfied with their services. This builds a longer-term relationship with clients and is often the key to future work.

You can build job security by keeping your employer satisfied with your work after you have been hired. A good way to do this is to develop and use a lifelong learning plan. In the future, you can keep your skills up to date by enrolling in technical classes, by attending workshops and seminars, by reading technical journals, and by active membership in professional organizations.

6.3 Why Capturing Information Is Important to You

Capturing information is an important part of developing and using your personal marketing plan. When you learn new concepts in your classes, you work hard to remember them, at least long enough to score well on exams. However, this information is important for more than receiving high grades. It sets the foundation for success in future classes and your working career.

Your experiences, activities, and skill development are a source of information to use in several ways in your career. Early classes provide tools and skill development as a prerequisite for following classes. In a similar way, early experiences in your working career are a foundation on which you can build more challenging and rewarding job activities.

When you are involved in various projects and activities, you may think you will remember details. Although this may be true for a short time, the more you learn and the greater number of experiences you have, the more difficult it becomes to remember them.

You can remember details better when you record them.

Your goal should be more than just recording information. The value of keeping records is to retrieve information about your experiences and skills. An information capturing system (ICS) is a structured method to record and retrieve information to help you build a successful career.

As you expand your knowledge, you can prevent information overload and retrieve information by developing an ICS. This is a systematic way for you to record your knowledge, skills, and experiences and then retrieve this information for a variety of uses. In this chapter, we show you how you can establish and use your information capturing system.

6.4 Establishing Your Information Capturing System

Your capturing information system is the foundation of your personal marketing plan as demonstrated in Figure 6.2. Let's see how you can learn to capture appropriate information so you can retrieve it when you want to use it for activities associated with your job search and other career job changes.

In establishing your capturing system, you have options for software and hardware. Select familiar software, such as MS Word, and format files for easy access. This will encourage you to use your capturing system, as indicated in the picture in Figure 6.3.

Let's think about the structure of your personal marketing plan. Most applications of stored information will be in word processing documents, so, it is logical to select the same word processing software for your capturing system. By doing this, you can save time by using the copy/paste functions to move items from storage to a new document.

Initially, you will use captured information to prepare resumes and cover letters. When starting your job search, you will use additional information about companies and how you can most effectively prepare for specific job interviews.

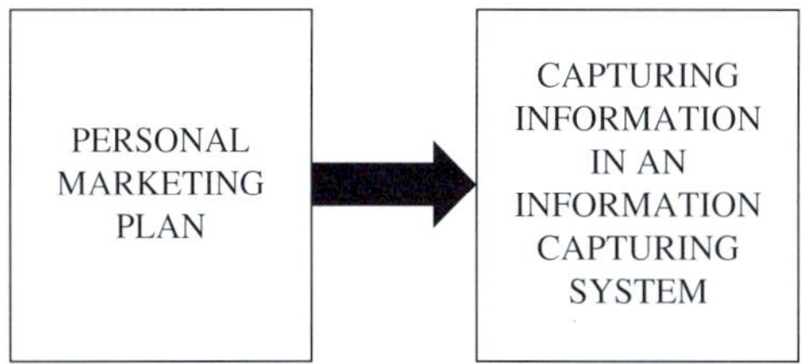

Figure 6.2
ICS capturing for personal marketing plans.

Figure 6.3
Information capturing system.

INFORMATION CAPTURING CATEGORIES

- Personal characteristics and skills
- Employment market
- Portfolio
- Job interview preparation
- Career history

Figure 6.4
Categories of information capturing systems.

Other information will be useful later in your working career when you apply for more challenging work positions, and when you are considered for recognition and awards. The easiest way to retrieve information is to record it in the sections shown in Figure 6.4.

Select details of the structure of your capturing system to fit your personal style of handling information. Remember, the most important part of your storage system is to design it so you can quickly and easily retrieve information when needed.

6.5 Using Your Information Capturing System

After you select software and format your capturing system, the next step is to record and use information, as shown in Figure 6.5. Although it may be early in your career, you already have several sources of information.

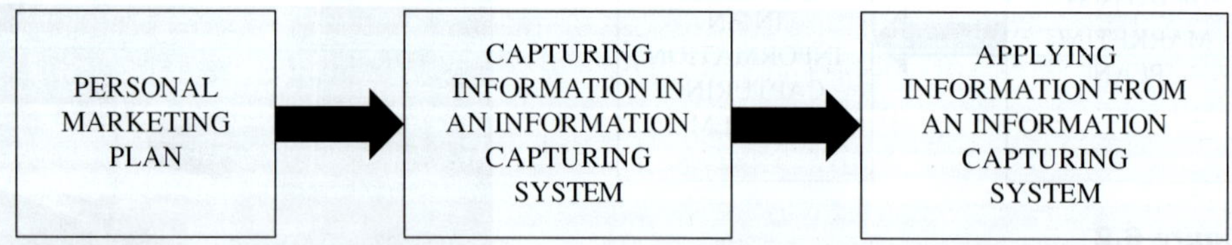

Figure 6.5
ICS process overview.

Let's review sources of information for each of the five sections of your ICS listed in Figure 6.4 and then discuss how you can apply this information in your job search and future career activities. This process is diagramed in Figure 6.6 to show how captured information can be used for different applications.

6.5.1 Personal Characteristics and Skills

Personal Characteristics and Skills—Information Source You are the best source for information for your personal characteristics and existing skills. Use the outline in Figure 6.7 to think about the characteristics that help you achieve success. Some of these

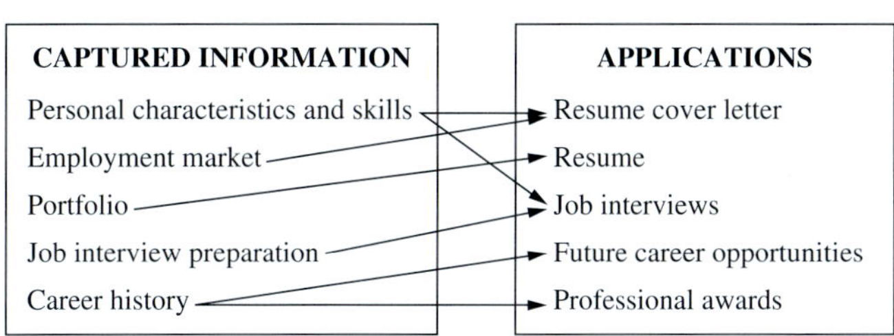

Figure 6.6
Application of information from ICS.

ICS SECTION 1
PERSONAL SKILLS

Outstanding personal characteristics
Educational experiences
  Classes and laboratories
  Projects
Personal development
  Problem solving skills
  Communication
  Leadership and teamwork skills
  Project management
Work experiences

Figure 6.7
ICS section 1.

may relate to how you learn and use new information. Others may be associated with your interpersonal skills. For an additional perspective, ask friends and relatives what they think are your outstanding personal characteristics and add these to this section.

> Identify your personal characteristics and special skills.

Personal Characteristics and Skills—Information Applications Information from the personal characteristics and skills section is an important resource to help you prepare effective cover letters for your resumes. In this chapter, we use the term cover letter to include letters and electronic cover messages.

Although some companies focus more on resumes and less on cover letters for information, it is still a good idea to include a cover letter or other cover message when sending your resume. Your cover letter is an opportunity to add information and discuss your qualifications for a specific position. Typical information in a cover letter is shown in Figure 6.8.

The purpose of your cover letter is to encourage potential employers to read your resume. A good way to do that is to identify details about the job opening and then tell them how you are uniquely qualified for that position. So, select information from your personal characteristics and skills section that makes your cover letter an effective step in your job search process. A sample cover letter is shown in Chapter resource 6.1.

Information from your personal characteristics and skills section can also be useful during job interviews. Remember, we compared your job interviews to sales calls, so "sell" your special skills by discussing your most important, and unique, characteristics during your job interviews.

6.5.2 Employment Market

Employment Market—Information Source In the previous section, we identified the value of discussing job opening details in your cover letter so you could tell a

RESUME COVER LETTER
Address specific person/title
Paragraph 1
Demonstrate your company research
Paragraph 2
State your major qualification
Tell how you are uniquely qualified
Paragraph 3
State additional qualifications
Paragraph 4
Request interview
Tell when you are available
Tell how you will follow-up

Figure 6.8
Cover letter content.

ICS SECTION 2 EMPLOYMENT RESEARCH
University Career Center
Workshops
Company profiles
Advisors
Professional Activities
Guest speakers
Networking
Field Trips
Career fairs

Figure 6.9
ICS section 2.

company how you are qualified for a specific position. How can you find these details? Figure 6.9 identifies sources of information about employment opportunities.

One of your best sources of details about job opportunities and other employment market information is your University Career Center, as indicated in Figure 6.10. These campus centers have information about companies and offer workshops and other resources to prepare you for successful job interviews. You can maximize benefits from the University Career Center when you establish contact early and take advantage of their resources.

You can also find information for your employment market section by reading about companies and potential positions for engineers. Think about engineering positions that interest you, search the Internet and other resources, and record information about companies who use engineers in these positions.

Figure 6.10
Career center resources.

Employment Market—Information Application Use information in this section that summarizes the results of your company research to select companies to contact. Then, you can tailor your cover letter to fit the particular company and job opening. Include statements that demonstrate your research of the company and what you can bring to them to help them improve their profitability.

Do your homework and research companies before contacting them.

6.5.3 Portfolio

Portfolio—Information Source As introduced in our discussion of your personal marketing plan, the portfolio section of your ICS is a description of your skills and experiences, as outlined in Figure 6.11. Chapter resource 6.2 is a good example of a typical portfolio.You have many resources to help you develop your portfolio, including your classes, laboratories, professional organizations, and other student activities.

"It is never too early to begin developing your network of contacts. Everyone you know has the potential to become an important friend and business ally. Some of my best contacts go back long ago to the college days when our only interests were passing a mid-term exam and making 'typical' college student weekend plans.

Building a good project resume is equally important. Whether it is to maintain your marketability or to use in building a business, keeping a portfolio of your work can help you stay in control of your future. Both a strong network of contacts and portfolio of projects were instrumental in building my companies."

(Jim Phillips, PE, Electrical Engineer, Founder of Brainfiller.com and ArcFlashForum.com)

ICS SECTION 3
PORTFOLIO

Personal contact information
Education Summary
Relevant courses
Technical skills
Work experiences
Professional activities
Honors and awards
List of completed courses and grades
- Major area of study
- Other related courses
- Non-technical courses

Figure 6.11
ICS section 3.

Your experiences are greatly expanded when you have coop and intern jobs. These positions provide opportunities for you to learn and apply new technical skills, and you can develop many nontechnical skills when you work with other engineers. Use the University Career Center to find part-time work positions early in your academic experiences, and then add the results of these experiences to the portfolio section of your ICS.

Your University Career Center will also help you prepare resumes. Although you may not use a resume for some time, it is much easier to prepare one if you take advantage of the University Career Center workshops and resume reviews now, and then add information you receive to your portfolio. An effective way to put information into your portfolio is to use a resume format.

Portfolio—Information Application Your portfolio is an important resource when preparing resumes for coop, intern, and after graduation positions. Because each resume should be developed for a specific job opportunity, it is good to build a large amount of information in your portfolio. Then, you have the option of selecting the most appropriate information for the specific job opportunity.

Use a resume format to structure your portfolio.

As we discussed, the primary purpose of your resume is to open the door for a personal interview. In addition to the captured information in your portfolio, your research of companies and engineering positions can be used to tailor your resume for each specific job opening.

A comparison of typical information for initial resumes and future resumes after graduation is shown in Figure 6.12. Personal contact and education sections are common to all resumes. Other resume sections, such as skills developed in various classes, laboratories, and in work experiences should be specific to the job under consideration.

Until you have several years of work experience, your resume should be limited to one page. However, when preparing resumes for coop and intern positions, you can add a second page and list completed courses and grades. This helps an employer select a job that fits your skills.

INITIAL RESUMES (Resume for coop and intern positions)	FUTURE RESUMES (Resume for after graduation positions)
Personal Contact information	Personal Contact information
Education	Education
Skills	Relevant Courses
Work Experience	Skills
Projects and Activities	Work Experience
Honors	Projects and Activities
Academic History	Honors

Figure 6.12
Different types of resumes.

Make each resume unique for a specific job opportunity.

Let's demonstrate how you can use information in a portfolio to prepare resumes. In the sample portfolio shown in Chapter resource 6.2, the format is similar to the format of a resume. The advantage of using the same format for both is that you can use the copy/paste tool to transfer information from the portfolio to resumes.

Initially, your portfolio will be limited to information and skills obtained during the early years of your undergraduate experiences. As you near graduation, your portfolio will contain more information than would be included in a single resume. However, the real value of a portfolio is the flexibility to select information for your resume that is most relevant to each job opening.

Chapter resource 6.3 shows a resume based on the portfolio of a student when applying for a coop position as a sophomore. Note how the portfolio resource was used in the following areas:

- Education (early GPA)
- Skills (some skills not yet available)
- Work experience (early work experiences)
- Projects and activities (early projects and activities)
- Honors (limited to early honors)
- Academic history (listed on second page)

Now, compare this to the sample resume for after graduation in Chapter resource 6.4, which includes

- Education (later GPA)
- Relevant courses (more appropriate than academic training)
- Skills (expanded to include new skills)
- Work experiences (coop experiences)
- Projects and activities (expanded involvement in projects, organizations, and activities)
- Honors (additional semesters on Dean's List and member of honor societies)

As these examples demonstrate, the portfolio is the source of information, and each resume should be prepared by selecting the most appropriate items for the specific job.

6.5.4 Job Interview

Job Interview—Information Source A list of appropriate information to record prior to job interviews is shown in Figure 6.13. An excellent source of information for this section is your University Career Center. The staff of this center provides workshops and other information to help you prepare for job interviews.

In addition to suggestions about how to successfully participate in job interviews, your University Career Center will also help you build confidence and gain experience by holding mock interviews. The key ideas you learn from literature, workshops, and mock interviews should be added to this section of your ICS.

ICS SECTION 4
JOB INTERVIEW

Before the Interview
- Potential questions to answer
- Questions to ask

Going to the Interview
- Appropriate dress
- Time, location, contacts

During the Interview
- Starting, ending
- Answering questions
- Asking questions and taking notes

Follow-up activities
- Sample letters/e-mail messages

Figure 6.13
ICS section 4.

ICS SECTION 5
CAREER HISTORY

Part-time, coop, and intern jobs
- Dates and contacts
- Major responsibilities
- Skills applied

Full-time positions
- Dates and contacts
- Major responsibilities
- Team projects
- Skills enhanced

Professional activities

Figure 6.14
ICS section 5.

Job Interview—Information Application Use information in this section to prepare for an interview. For example, you can make a list of questions to ask during the interview by reading sample questions stored in this section. Your review of information you obtained from the University Career Center helps you be more confident to successfully participate in the interview.

Because much of the information in this section was obtained from workshops and mock interviews, you can use this to help you remember the important points to include in actual job interviews. A reasonable amount of time to review this information will pay big dividends in increasing your success to receive job offers.

6.5.5 Career History

Career History—Information Source Start this section of your ICS by recording information from coop, intern, and other part-time jobs. In addition, record your experiences in senior design team projects and professional organization projects as part of your career history. Typical information in the career history section of your ICS is shown in Figure 6.14.

After you graduate, this section continues to become more valuable each time you record information relating to your work experiences and professional activities. When you learn new skills, participate in workshops, and expand your ability to be ready for new challenges, you should add this information to this section of your ICS.

Career History—Information Application Your career history is an excellent resource to help you publicize your skills so others recognize your contribution to your employer. This typically would be associated with work after graduation, but could also help you obtain more opportunities for experience in part-time work while you are in school. More importantly, you can use information about what you have accomplished if you need to change jobs.

> ***"In every Life Science proposal we send to our customers we include employee profiles or CVs (curriculum vitae). Many times these profiles are required by the customer and can be the deciding factor if we are awarded the project or a competitor is. Maintaining your educational and professional history in a documented form will assist you and your company when an opportunity appears.***
>
> ***Maintaining this history will also be of value when your annual review is due and your manager solicits your feedback about what you accomplished in the last year. Having this information in an organized and detailed manner may be all you need to impress your manager and others critical to your professional development whether it's a new role, position or job opportunity."***
>
> **(Chris Phillips, PE, EET, and Northeast Regional Manager at RoviSys)**

As indicated in the opening questions of this chapter, your world is very dynamic. When companies experience change, you may be able to benefit from that change when you use information in your career history to publicize your skills. If your present job disappears or your employer is no longer in business, you can use your career history to improve your success in finding the next job.

You may also initiate career path changes based on the dynamics of technology. When you plan for the future and develop new skills, you will be better prepared for unexpected opportunities. An interesting outcome of your ICS is that it encourages you to gain new skills and add these to your career history.

Your career history section is also a valuable resource of information to use to receive awards for outstanding work. Often, awards are given by your company and peers in your profession. In addition to recognizing your good work, awards also encourage others to reach higher levels of achievement. So, there can be many benefits of building your career history record.

We end our discussion of this chapter by revisiting our opening questions and seeing how one professional engineer uses information capturing on the job to achieve success in this very dynamic and challenging world. The following conversation is with Elmer Paine, an electrical engineer who has worked for a Fortune 100 company and several consulting firms.

A CONVERSATION WITH ELMER PAINE

Elmer, you raised some very interesting questions in the information capturing scenario at the beginning of this chapter. As a consulting engineer, why do you capture information and keep records of your activities?

Answer

"There are many good reasons to capture and maintain a record of my professional experience history. The obvious reason is maintaining my employment history. Beyond capturing and maintaining details of my previous employment situations, I have always tried to maintain some level of contact with as many of my previous supervisors and mentors as possible.

"In addition to the more common practice of keeping a resume of employment positions throughout my career, I have found it very useful to maintain a daily journal of what I do. I also keep an up-to-date list of clients and projects.

"There is a very practical reason for maintaining a daily professional diary/journal and a comprehensive list of projects. As a practicing consulting engineer, I am called upon to Seal and Sign construction plans and specifications I produce as instruments of my design work. These documents are contract documents, and there is a great deal of personal liability that I accept every time I stamp and sign a set of plans and specifications for a construction project.

"There are no statute of limitations on the Engineer's personal liability and, over the course of his or her career, there will possibly be hundreds or thousands of documents bearing his or her stamp. Therefore, it is incumbent upon the Registered Professional Engineer to keep very detailed project records of every project that bears his or her Seal."

So, how do you record and use information?

Answer

"In the consulting engineering profession, I am always interviewing with clients for the next project design commission. Even with clients for whom we have previously worked, we are frequently required to prepare and submit 'qualifications,' a part of the procurement process the client uses to hire design professionals. This is always the case when getting hired for publically funded government projects such as universities, Veterans' hospitals, and Courthouses.

"The qualifications submittals are typically required to list from three to ten projects along with pertinent details about each that are similar in type and scope with the project for which the client is hiring his or her design team. Having an up-to-date list of prior projects is essential because there is very limited time between when the Request for Qualifications (RFQ) is published and the deadlines when all prospective design teams are required to submit their proposals. There is never enough time to recreate the list every time an RFQ is published.

"I frequently update my list. It is best to do so immediately after completion of every project/assignment while the details are still fresh in my mind and the project documentation is still readily available prior to its being archived off site. Additionally, I categorize my project list by type (i.e., hospitals, performing arts venues, cultural arts, courthouses, data centers, commercial, office buildings, higher education facilities, airports, etc.) and by design discipline (medium voltage systems design, life safety systems design, lighting and controls, power distribution design, fire alarm, nurse call, emergency generators, etc.)."

What nontechnical skills do you use in your consulting work?

Answer

"The nontechnical skills necessary for capturing and documenting my project experiences include the self-discipline (perseverance) over many years to maintain a project experience database. I use Microsoft Excel to log my projects so I can quickly sort the list for particular type of project, disciplines included, and reference contacts. In addition, I maintain and nurture personal relationships with past supervisors, mentors, previous clients, and prospective employers so when it is necessary for a client or prospective employer to validate my history and/or experience, it can be accomplished with very little effort."

What are the most important benefits of capturing information?

Answer

"Resources, such as past supervisors and mentors as references, have been necessary for me to obtain professional registrations in many states across the US. In 30 years of professional practice, I have had the opportunity for fifteen different employment assignments and/or employers each having a different 'boss.' Each and every one of those opportunities was aided by my ability to present evidence of my experience and portfolio.

"After almost 17 years in the consulting business, my project experience list has over 300 separate projects in about ten project categories. Obviously, it would be impossible to remember all of those projects without gathering the data in real time along the way. When my firm receives an invitation to present a qualifications package for a proposed design team for an upcoming project, it seldom takes me more than ten minutes to assemble a list of representative projects from my overall body of work during the past 17 years.

Of course, when the opportunity (necessity) to change employers manifests itself, it has always been very impressive to be able to show a prospective new employer a document showing a comprehensive listing of my experience and body of work."

What advice would you share with students?

Answer

"The quality of work engineers do must, at a minimum, conform to the "standard of care" for their profession. This is the big leagues and errors result in wasted time, wasted money, and sometimes death. The young engineer needs to start very early and develop the HABIT of taking notes throughout the day. His or her notes could help avoid a costly damage claim or litigation. Also, detailed project records are necessary for every engineer throughout his or her career."

(Elmer Paine, PE, Electrical Engineer, and Sr. Electrical Engineer at Affiliated Engineers)

6.6 Conclusion

Successful engineers develop the ability to think logically and to apply this approach to nontechnical areas of their career. As an engineering student, you can apply this same process to capture, retrieve, and use important career related information.

The University Career Center is an excellent source of information and training. You can use their resources when developing job related parts of your ICS. This will save you time and will prove to be a valuable asset when seeking jobs as a student and after graduation.

Your information capturing system should be dynamic and used to capture experiences during your lifetime career. Learning how to use this tool will pay you big dividends now and throughout your career. So, take advantage of the opportunity to initiate your ICS now, and you will enjoy many benefits over a long period of time.

END OF CHAPTER REVIEW QUIZ

Select the most appropriate answer to the following statements.

1. It is important for you to capture information
- **a.** To use as a resource to prepare future resumes.
- **b.** To prepare for more successful job interviews.
- **c.** To have a record of achievements and be ready for unexpected job opportunities.
- **d.** To accomplish all of the above.

2. The major objective of your cover letter is to
- **a.** Make sure your resume is delivered to the correct person.
- **b.** Encourage others to read your resume.
- **c.** Be invited to a job interview.
- **d.** Identify you are the author of your resume.

3. The primary use of a well constructed portfolio is
- **a.** To save time and be an effective resource when preparing resumes.
- **b.** To show your parents and friends what you have accomplished.
- **c.** To prepare cover letters and e-mails when sending your resume.
- **d.** To provide information when preparing for a job interview.

4. The main purpose of your resume is to
- **a.** Get a job after graduation.
- **b.** Take to job fairs so you can show prospective employers all your accomplishments.
- **c.** Be invited to an interview with a recruiter from a company.
- **d.** Be a document of your education and work experiences.

5. Information in your personal characteristics section of your ICS is a resource to
- **a.** Include in a cover letter.
- **b.** Make your cover letter specific to the job opening requirements.
- **c.** Remind you of some items to discuss in a job interview.
- **d.** Accomplish all of the above.

6. Your University Career Center provides
- **a.** Information about prospective employers.
- **b.** Workshops on resumes and cover letters.
- **c.** Mock interviews.
- **d.** All of the above.

7. The most effective software for information capturing is
   a. Available from the Web.
   b. A very detailed system so you can record everything you do.
   c. One that you can put on a wireless device.
   d. One that fits your personal style of handling information and is easy to use.
8. The most important value of coop, intern, and part-time work is
   a. The experiences that you can add to your portfolio.
   b. The money you can earn while going to school.
   c. The opportunity to not have classes every semester.
   d. None of the above.
9. Your information capturing process should
   a. Be initiated when you start your senior year when you have a lot of information to record.
   b. Be started now and be maintained by periodically adding appropriate information.
   c. Be copied from a friend so you don't have to spend much time to start using it.
   d. Be updated every day to be effective.
10. The career history section of your information capturing system is
   a. A record of your experiences and successful activities as a student.
   b. Used primarily to request awards and special recognition.
   c. A good resource when you consider major changes in your career path.
   d. Important to show your employer why you should be promoted.

EXERCISES TO DEVELOP AND ENHANCE YOUR SKILL SET

Exercise 6.1 Develop an information capturing system, prepare a written report based on the following outline, and deliver your report to your instructor at the start of the class session as assigned. Send an electronic copy of your report to your instructor.

Information Capturing System (ICS) Report

Prepared by ________________________________

Overview of Information Capturing System

Discuss the system you selected or enhanced to capture information.

Explain why you selected this system and how you will retrieve information.

Information Capturing System Structure

Provide an outline of the major sections of your system.

Explain why you selected these sections and the sources of information to be added to each section.

University Career Center Resources

Discuss the resources available at the University Career Center.

Explain how you plan to use these resources and how information obtained from this center will be applied to your information capturing system.

Information Capturing Process Assessment

Report your assessment of the value of using an information capturing system. Rate your experience from 1 to 10 (10 being the highest value) and give reasons for your numerical rating. Discuss how you will apply what you learned in this experience to future classes and your working career.

END OF CHAPTER RESOURCES

Chapter resource 6.1

Sample Cover Letter

1234 Lake Avenue
Cleveland, OH 44102-1234

Date:

Mr. Thomas Edison
Manager of Operations
Ohio ABC Company
1234 Main Street
Norka, Ohio 44431

Dear Mr. Edison:

When Jean Snyder of your Human Resource Department spoke to me today, she indicated that you have an opening for an Associate Engineer. As a result of our discussion, I believe my background and experience will benefit Ohio ABC Company.

This spring, I will receive a Bachelor of Science degree in Electrical Engineering from Cleveland State University. In addition to my courses in engineering, I participated in a Senior Design Team Project to determine how computer software can be applied to schedule electrical equipment maintenance. During my four years in college, I maintained a 3.5 GPA while carrying a full schedule of course work and working part time in the computer laboratory.

My summer work experience at MTA provided a good introduction to the operations of a large electrical system. As indicated in the attached resume, I also developed oral presentation skills in my project team assignment. In addition, I was the chair of the publicity committee for a major conference for students last fall. These experiences, along with my academic background, should be a valuable asset to your company.

I am interested in the position of Associate Engineer and am requesting an interview to discuss how my education and work experiences can enhance your company's operations. I will call next Tuesday to discuss when we can meet.

Sincerely,

Mary Ann Brown

Mary Ann Brown

Chapter resource 6.2

Sample Portfolio

Education

Fenn College of Engineering, Cleveland State University, Cleveland, Ohio
Bachelor of Science in Electrical Engineering
Expected Graduation Date: May 2013
Early Overall GPA 3.23/4.00; Major GPA 3.55/4.00
Later Overall GPA 3.39/4.00; Major GPA 3.67/4.00

Relevant Courses

Power Systems
Power Electronics & Lab
Control Systems & Lab
Electromechanical Energy Conversion
Communications & Lab
Engineering Management
Industrial Safety and Health

Skills

- Programming skills: C++ and Java
- Functional programs: MATLAB, Pspice, Solid Works
- PLC programming

Work Experience

9/11	The Electric Utility Company, Athens, Ohio
to	Customer Service (Co-op Position)
12/11	Worked with Senior Engineer to survey customer power uses and requirements
	Recommended methods to reduce power demand and cost with improved results
	Provided power use calculations and reports for customer
1/11	The Best Engineering Consultants, Cleveland, Ohio
to	Draftsperson (Co-op Position) and member of Power Station Design team
5/11	Provided drawings and wiring diagrams of electrical and system protection equipment
	Used AutoCAD and AutoCAD inventor
6/10	The Any Motor Company, Detroit, Michigan
to	Technician A
9/10	Maintained electronic controls and applied changes to control systems
	Analyzed problems on programmable controllers on automation equipment
6/09	Fast Food Company, Cleveland, Ohio
to	Assistant Manager (Part time)

9/09 Directed activities of four employees
Operated cash register and provided customer service

Projects and Activities

Member of Science Project to design and build a robot
Learned how to combine electronics, mechanics, and sensors to perform functions
Participated in team design of robot that was able to pick and place objects

Member of Mini Baja Vehicle team at CSU
Worked with team leader on ignition system
Designed electronic ignition system
Researched systems and applied principles to project; vehicle met performance objectives

Senior Project: Designed control system for a wind turbine generator with senior design team

Professional Organizations: IEEE (Treasurer), Tau Beta Pi, Eta Kappa Nu

Activities: Captain of Golf Team, Toastmasters Club

Honors

- Northeast Ohio Science Fair 2nd Place Award
- National Merit Scholarship and Cleveland State University Scholarship
- Dean's List four semesters
- Tau Beta Pi and Eta Kappa Nu

Academic History

Major Discipline	
Electric Circuits I	A
Electronics I	A
Electric Circuits II	B
Electronics II	A

Other Engineering Subjects	
Engineering Statistics and Probabilities	A
Project Economics	A
Thermodynamics	B

Math and Science	
Chemistry I	B
Physics I	A
Calculus I	A
Physics II	A
Calculus II	B

Nontechnical	
Western Civilization	A
Latin History	C
English I	B
Speech	B
Urban Studies	B

Chapter resource 6.3

Sample Resume for Co-op and Intern Positions

(Note: the following information is on the first page)

JOHN Q. SMITH

1980 East 24th Street (University)	216.687.xxxx
Cleveland, Ohio 44115-2425	j.smith@whereever.com
9755 Some Street (Home)	440.625.xxxx
Mentor, Ohio 44060	

Education

Fenn College of Engineering, Cleveland State University, Cleveland, Ohio
Electrical Engineering
Expected Graduation Date: May 2013
Overall GPA 3.23/4.00; Major GPA 3.55/4.00

Skills

- Functional programs: MATLAB, Pspice, Solid Works
- PLC programming

Work Experience

6/10 to 9/10	The Any Motor Company, Detroit, Michigan Technician A Maintained electronic controls and applied changes to control systems Analyzed problems on programmable controllers on automation equipment
6/09 to 9/09	Fast Food Company, Cleveland, Ohio Assistant Manager (Part time) Directed activities of four employees Operated cash register and provided customer service

Projects and Activities

Member of Science Project to design and build a robot
  Learned how to combine electronics, mechanics and sensors to perform functions
  Participated in team design of robot that was able to pick and place objects
Member of Mini Baja Vehicle team at CSU
  Worked with team leader on ignition system
Professional Organizations: IEEE

Activities: Golf Team

Honors

- Northeast Ohio Science Fair 2nd Place Award
- National Merit Scholarship and Cleveland State University Scholarship
- Dean's List one semester

(Note: the following information is on the second page)

JOHN Q. SMITH

1980 East 24th Street (University)	216.687.xxxx
Cleveland, Ohio 44115-2425	j.smith@whereever.com
9755 Some Street (Home)	440.625.xxxx
Mentor, Ohio 44060	

Academic History

Fenn College of Engineering
Cleveland State University
Electrical Engineering

Academic History

Major Discipline		Other Engineering Subjects	
Electric Circuits I	A	Engineering Statistics and Probabilities	A
Electronics I	A	Project Economics	A
Electric Circuits II	B	Thermodynamics	B
Electronics II	A		

Math and Science		Nontechnical	
Chemistry I	B	Western Civilization	A
Physics I	A	Latin History	C
Calculus I	A	English I	B
Physics II	A	Speech	B
Calculus II	B	Urban Studies	B

Chapter resource 6.4

Resume for After Graduation

JOHN Q. SMITH

1980 East 24th Street (University)	216.687.xxxx
Cleveland, Ohio 44115-2425	j.smith@whereever.com
9755 Some Street (Home)	440.625.xxxx
Mentor, Ohio 44060	

Education

Fenn College of Engineering, Cleveland State University, Cleveland, Ohio
Bachelor of Science in Electrical Engineering
Expected Graduation Date: May 2013
Overall GPA 3.39/4.00; Major GPA 3.67/4.00

Relevant Courses

Power Systems	Communications & Lab
Power Electronics & Lab	Engineering Management
Control Systems & Lab	Industrial Safety and Health
Electromechanical Energy Conversion	

Skills

Programming skills: C++ and Java
Functional programs: MATLAB, Pspice, Solid Works
PLC programming

Work Experience

9/11	The Electric Utility Company, Athens, Ohio
to	Customer Service (Co-op Position)
12/11	Worked with Senior Engineer to survey customer power uses and requirements
	Recommended methods to reduce power demand and cost with improved results
	Provided power use calculations and reports for customer
1/11	The Best Engineering Consultants, Cleveland, Ohio
to	Draftsperson (Co-op Position) and member of Power Station Design team
5/11	Provided drawings and wiring diagrams of electrical and system protection equipment
	Used autoCAD and autoCAD inventor

Projects and Activities

Member of Mini Baja Vehicle team at CSU
Designed electronic ignition system
Researched systems and applied principles to project; vehicle met performance objectives

Senior Project: Designed control system for a wind turbine generator with senior design team

Professional Organizations: IEEE (Treasurer), Tau Beta Pi, Eta Kappa Nu

Activities: Captain of Golf Team, Toastmasters Club

Honors

- Dean's List six semesters
- Tau Beta Pi and Eta Kappa Nu

7 Engineering Disciplines—A Fantastic Way to Have Fun!

"Engineers are dreamers who have the ability to make those dreams come true!"

—Charles Alexander

7.1 Introduction

The good news is that regardless of what is happening with world economies, engineering jobs tend to always be there. Sometimes it requires more effort to find them and sometimes one has to move to where the jobs are. In addition, of the top ten paying jobs today available to undergraduates, eight are in engineering (including computer science), see Figure 7.1. The seventh highest paying job is in physics and the eighth is in applied mathematics.

Historically engineers have received twenty job offers for every one received by a nonengineering graduate. This holds true especially during periods of economic uncertainty.

Engineering graduates have historically received some of the highest starting salaries among graduates especially for graduates who receive master's degrees!

As an undergraduate, you will need to pick a specific discipline in which to earn your first degree. The intent of this chapter is to make that selection process easier for you. We start by looking at what drivers will create engineering jobs in the future. We have identified five major drivers of engineering employment in the twenty-first century. There will be others, but we feel these will be the major ones. In addition, detailed responses to each of these areas and how they will develop could demand chapters by themselves and in some cases whole books. We only briefly touch on each of them and only from the standpoint of what types of engineers will be needed.

LEARNING OBJECTIVES

By using the information and exercises in this chapter you will be able to:

- Understand the multitude of engineering disciplines that are available.
- Understand the incredible breadth and depth of careers available to engineers.
- Understand the growth areas in the world economy that will support engineering jobs.
- Know where to access additional information regarding specific engineering disciplines.
- Identify the various professional organizations that support specific engineering disciplines.
- Understand that engineering is one of the most exciting careers available to anyone today!
- Finally understand that a career in engineering will be "fun!"

	Ten highest-paying degrees (undergraduate) Major	Median starting salary	Mid-career average
1	Petroleum engineering	$97,900	$155,000
2	Chemical engineering	$64,500	$109,000
3	Electrical engineering	$61,300	$103,000
4	Materials science and engineering	$60,400	$103,000
5	Aerospace engineering	$60,700	$102,000
6	Computer engineering	$61,800	$101,000
7	Physics	$49,800	$101,000
8	Applied mathematics	$52,600	$98,600
9	Computer science	$56,600	$97,900
10	Nuclear engineering	$65,100	$97,800

Figure 7.1
Ten highest-paying degrees.
Source: Jobs.aol.com and OnlineDegrees.com

The five major drivers for engineering employment in the twenty-first century are

- Manufacturing
- Affordable, high-quality healthcare
- Revitalizing and enhancing the infrastructure
- Energy independence and economic stability
- Enhancing the environment

A lot can be learned about specific engineering disciplines by going to the various websites of the different engineering professional societies. We will identify many of these so you can learn more.

7.2 Engineering Skills Needed for "Manufacturing"

> We will see manufacturing flowing into those countries that can support highly automated manufacturing facilities with highly skilled labor.

During the end of the twentieth century and the beginning of the twenty-first century, manufacturing facilities mostly followed the lowest cost labor. However, that trend will be short lived; we will see manufacturing flowing into those countries that can support highly automated manufacturing facilities with highly skilled labor. These facilities will require almost all fields of engineering including but not limited to chemical, civil, computer, electrical, mechanical, and software engineering.

EXAMPLE 7.2.1

Expand the list of engineering disciplines associated with manufacturing. We start by performing an Internet search using Bing on "manufacturing engineering disciplines." This produces an extensive list such as shown in Figure 7.2.

PRACTICE PROBLEM 7.2.1

Use the Internet to further expand the list of engineering disciplines associated with manufacturing from those found in Example 7.2.1.

7.3 Engineering Skills Needed for "Affordable, High-Quality Healthcare"

The healthcare systems currently in use in every country are inadequate to supply consistent healthcare and high-quality healthcare for a reasonable price. So, can such a system be developed? Fortunately the answer is yes! It will primarily fall on the engineering community working closely with medical professionals to develop the technologies to make high-quality healthcare widely available at a reasonable cost.

> Affordable, widely available, high-quality healthcare can only come from the engineering community working closely with medical professionals.

1-10 of 21,100,000 results· Advanced

Manufacturing engineering

The **Engineering** Alphabet: An Overview of **EngineeringDisciplines**. Careers in **engineering** span the alphabet. From aerospace to **manufacturing** to transportation ...
www.sinc.sunysb.edu/Stu/hnaseer/photo.htm

Overview of EngineeringDisciplines:

• The "Medium Four" **Disciplines**: Aerospace, Biomedical, Chemical, and Industrial/**ManufacturingEngineering**, which collectively account for approximately 20% of all ...
www.dedicated **engineers**.org/Resources/**Engineering_Disciplines** ... · PDF file

About EngineeringDisciplines **- The Online Guide to** Engineering **...**

The "Medium Four" **Disciplines**: Aerospace, Biomedical, Chemical, and Industrial/**ManufacturingEngineering**, which collectively account for approximately 20% of ...
www.find**engineering**schools.org/Resources/about_**disciplines**.htm

Related Searches for manufacturing engineering disciplines

Chemical **engineering** ·
Civil **engineering** ·
Electrical **engineering**

Engineering is the **discipline**, art, and profession that applies scientific ... the production process of paper products; Molecular **engineering**: Focuses on the **manufacturing** ...
en.wikipedia.org/wiki/List_of_**engineering**_branches

Engineering **- key** disciplines **|** Engineering **and** Manufacturing **...**

Engineering - key **disciplines**. This section focuses on key **engineering** roles within some of the main areas in the **engineering** and **manufacturing** sector.
www.ed.ac.uk/.../careers/explore/occupations/**engineer**/key-**disciplines**

Careers in ManufacturingTechnology **| eHow.com**

Manufacturing technology is a career **discipline** that focuses on design aspects of production equipment and **manufactured** goods. These careers have increased in recent ...
www.ehow.com/list_5949135_careers-**manufacturing-technology**.html

About the School's EngineeringDisciplines **| Mechanical and ...**

Figure 7.2
A selection from performing a search using Bing on the phrase "manufacturing engineering disciplines."

Biomedical engineering will be one of the most prominent if not the most prominent type of engineering to contribute to this area. In addition, electrical, computer, software, mechanical, chemical, and system engineers will be needed.

EXAMPLE 7.3.1

Expand the list of engineering disciplines associated with healthcare technology. We start by performing an Internet search on "healthcare engineering disciplines." This produces an extensive list, a small portion of which is shown in Figure 7.3.

1-10 of 25,800,000 results · Advanced

Healthcare Disciplines

Access to all **healthcaredisciplines** and jobs-Kelly **Healthcare** Resources is a global ... **EngineeringDisciplines**; Contact Us; Finance & Accounting. Finance & Accounting **Disciplines**

www.kellyservices.us/.../**Healthcare-Disciplines**

KSHE-Kentucky Society of HealthcareEngineers

Healthcare engineering disciplines include managing, operating and maintaining the physical plant facilities, communication and biomedical equipment and systems in ... www.kshe.org

Category: Engineeringdisciplines **- Wikipedia, the free encyclopedia**

Pages in category "**Engineeringdisciplines**" The following 78 pages are in this ... Management **engineering** (**healthcare**) Methods **engineering**; Microfactory; Molecular **engineering**

en.wikipedia.org/wiki/Category:**Engineering_disciplines**

Healthcare Engineering **(HcE) - College of** Engineering**, Purdue ...**

Healthcare Engineering (HcE) - College of **Engineering**, Purdue University ... A wide range of partnerships have been established with **healthcarediscipline** ...

https://**engineering**.purdue.edu/Engr/Research/Initiatives/HE

Related Searches for healthcare engineering disciplines

Figure 7.3
A selection from performing a search using Bing on the phrase "healthcare engineering disciplines."

PRACTICE PROBLEM 7.3.1

Use the Internet to further expand the list of engineering disciplines associated with healthcare from those found in Example 7.3.1.

7.4 Engineering Skills Needed for "Revitalizing and Enhancing the Infrastructure"

Another huge need for engineers will be in the area of revitalizing and enhancing our infrastructure. What we call the infrastructure includes things such as highways, bridges, water treatment and distribution facilities, the grid, power plants, buildings, and sewers. In addition, enhancing these really means things like smart highways, smart grid, and safer work zones in highways.

Revitalizing and enhancing our infrastructure will require a huge number of engineers!

Clearly civil engineers will be one of the most prominent if not the most prominent engineers working on these problems. In addition, we will need mechanical, electrical, computer, software, chemical, and system engineers.

PRACTICE PROBLEM 7.4.1

Use the Internet to expand the list of engineering disciplines associated with the infrastructure.

7.5 Engineering Skills Needed for "Energy Independence and Economic Stability"

Energy will probably be the where we will see the largest growth in both investment in dollars as well as investment in people (most of whom will be engineers) in the twenty-first century. If we look at the energy resources available to us, we can obtain needed energy from oil, gas, and coal as well as nuclear, hydro, solar, wind, geothermal, and tidal sources. The generation of electrical power will most likely be where we will see the greatest growth, and transportation, because it will increasingly be electrical, will be an important part of this. Some exciting possibilities can come from what may be the next-generation nuclear energy, liquid fluoride thorium reactors or LFTR as they are called (http://thoriumenergy.blogspot.com/). Also, using such variable alternative energy sources as wind and solar will require significant storage facilities such as flywheels, batteries, pumped hydro, and compressed air.

Since this will be such a large market, those countries that dedicate themselves to making these systems will generate significant revenues.Those countries that have energy resources for sale will also have positive cash flows.

Designing and building energy systems will require large investments in economic and human capital. Much of the human capital will need to be engineers.

PRACTICE PROBLEM 7.5.1

Use the Internet to expand the list of engineering disciplines associated with the electric power generation and resource development.

7.6 Engineering Skills Needed for "Enhancing the Environment"

If we look at our environment today we see much that needs to be fixed. It is also difficult to see how this can be done. However, if we look back 100 to 150 years ago we can see that technology can indeed fix our problems! Even our life spans have been increased by thirty-five years through technology (mostly from clean drinking water and modern sanitation facilities).

We will only have the kind of environment we want through the use of technology and engineering!

Engineers will continue to be at the center of improving our environment to whatever level we wish. Environmental engineers will obviously be needed and will be supported by chemical, civil, computer, electrical, mechanical, software, and systems engineers.

PRACTICE PROBLEM 7.6.1

Use the Internet to expand the list of engineering disciplines associated with enhancing the quality of our environment.

7.7 Engineering Professional Societies

A number of engineering societies can help you understand the various disciplines. Typically, these are the three major societies: the American Society of Civil Engineers (ASCE, http://www.asce.org/), the American Society of Mechanical Engineers (ASME, http://www.asme.org/), and the Institute of Electrical and Electronics Engineers (IEEE, http://www.ieee.org/). An expanded list can be found by going to the membership list of the American Association of Engineering Societies (AAES, http://www.aaes.org/international/). The list of member societies of AAES is shown in Figure 7.4 along with their web addresses. This list is by no means exhaustive.

The best way to find out what an engineering discipline does is to go to the various engineering society websites and explore them.

PRACTICE PROBLEM 7.7.1

Use the Internet to expand the list of engineering societies and organizations from the list shown in Figure 7.4.

In this chapter, wc use the professional societies to help you better understand what the various engineering disciplines do. In our chapter on the survey of engineering disciplines, you will see that your engineering societies are your most valuable asset in developing a network that will be absolutely essential for career development and technological growth!

Now let us look at three of the societies just mentioned.

7.8 The American Society of Civil Engineers, ASCE

Like all the larger as well as the smaller societies, ASCE is actually made up of several subdisciplines. These are presented in Figure 7.5. Note the wide variety of subdiscipline areas within the ASCE. The term civil engineering actually comes from civilian engineers, because up until this discipline was established almost all engineers were in the military. All the engineering disciplines we know today actually grew out of civil engineering. The web address for ASCE is http://www.asce.org/.

Name of the organization/location	Web address
American Association of Engineering Societies Washington, DC	http://www.aaes.org/international/
American Institute of Aeronautics and Astronautics AIAA Reston, VA	http://www.aiaa.org/
American Institute of Chemical Engineers AIChE New York	http://www.aiche.org/
American Institute of Mining, Metallurgical, and Petroleum Engineers, Inc. AIME Littleton, CO	http://www.aimeny.org/
American Institute of Medical and Biological Engineers AIMBE Washington, DC	http://www.aimbe.org/
American Nuclear Society ANS LaGrange Park, IL	http://www.new.ans.org/
American Society for Engineering Education ASEE Washington, DC	http://www.asee.org/
American Society of Civil Engineers ASCE Reston, VA	http://www.asce.org/
American Society of Mechanical Engineers ASME New York	http://www.asme.org/
Human Factors and Ergonomics Society HFES Santa Monica, CA	http://www.hfes.org/web/Default.aspx
Instituteof Electrical and Electronics Engineers - USA IEEE-USA Washington, DC	http://www.ieeeusa.org/
Junior Engineering Technical Society JETS Alexandria, VA	http://www.jets.org/
National Council of Examiners for Engineering and Surveying NCEES Clemson, SC	http://www.ncees.org/
National Society of Professional Engineers NSPE Alexandria, VA	http://www.nspe.org/index.html
Society of Women Engineers SWE Chicago, IL	http://societyofwomenengineers.swe.org/

Figure 7.4
Member engineering societies that comprise the American Association of Engineering Societies (AAES).

Civil engineering is actually composed of a multitude of subdisciplines such as environmental, structural, transportation, and water resource engineering.

PRACTICE PROBLEM 7.8.1

Go to the ASCE website and select two or three subdisciplines to determine what they do as practicing professionals.

Institutes, divisions, and councils of the ASCE
Architectural Engineering Institute
Coasts, Oceans, Ports, and Rivers Institute
Construction Institute
Engineering Mechanics Institute
Environmental & Water Resources Institute
Geo-Institute
Structural Engineering Institute
Transportation & Development Institute
Aerospace Division
Technical Council on Cold Region Engineering
Technical Council on Computing and Information Technology
Council on Disaster Risk Management
Energy Division
Technical Council on Forensic Engineering
Geomatics Division
Technical Council on Lifeline Earthquake Engineering
Pipeline Division
Technical Council on Wind Engineering

Figure 7.5
The institutes, divisions, and councils of the ASCE.

7.9 The American Society of Mechanical Engineers, ASME

Like civil engineering, mechanical engineering has a wide variety of subdisciplines and can be further explored at the ASME website, http://www.asme.org/. The various institutes, groups, and divisions of the ASME are listed in Figure 7.6.

Mechanical engineering is actually composed of a multitude of subdisciplines such as thermodynamic, biomedical, combustion, solar, wind, mechatronic, and control engineering.

PRACTICE PROBLEM 7.9.1

Go to the ASME website and select two or three subdisciplines to determine what they do as practicing professionals.

Institutes, Groups, and Divisions of the ASME	
INSTITUTES	International Gas Turbine Institute
	International Petroleum Technology Institute
	Nanotechnology Institute
BASIC ENGINEERING TECHNICAL GROUP	Applied Mechanics
	Bioengineering
	Fluids Engineering
	Heat Transfer
	Materials
	Tribology
ENERGY CONVERSION GROUP	Internal Combustion Engine
	Nuclear Engineering
	Power
	Advanced Energy Systems
	Solar Energy
ENGINEERING & TECHNOLOGY MANAGEMENT GROUP	Management
	Safety Engineering & Risk Analysis
	Technology & Society
ENVIRONMENT & TRANSPORTATION GROUP	Aerospace
	Environmental Engineering
	Noise Control Acoustics
	Rail Transportation
	Materials & Energy Recovery
MANUFACTURING TECHNICAL GROUP	Manufacturing Engineering
	Materials Handling Engineering
	Plant Engineering & Maintenance
	Process Industries
PRESSURE TECHNOLOGY GROUP	Nondestructive Evaluation
	Pressure Vessels & Piping
SYSTEM & DESIGN GROUP	Computers & Information Engineering
	Design Engineering
	Dynamic Systems & Control
	Electronic & Photonic Packaging
	Fluid Power Systems & Technology
	Information Storage & Processing Systems
	Microelectromechanical Systems

Figure 7.6
Institutes, groups, and divisions of the ASME.

7.10 The Institute of Electrical and Electronics Engineers, IEEE

The world's largest professional, technical society is IEEE. Currently it has over 400,000 members worldwide. In addition, it publishes approximately one-third of the world's literature in electrotechnology. Like ASCE and ASME, it has a large number of subdisciplines, which can be found at its website, http://www.ieee.org/. The IEEE societies are listed in Figure 7.7, and the IEEE councils are listed in Figure 7.8.

IEEE is the world's largest and most active technical, professional society and publishes approximately one-third of the world's literature in electrotechnology.

PRACTICE PROBLEM 7.10.1

Go to the IEEE website and select two or three societies and/or councils to determine what they do as practicing professionals.

7.11 Accredited Engineering Programs

If you still are looking for what program you may wish to select as your major, we recommend going to the ABET website (http://www.abet.org/). Figure 7.9 lists the different engineering disciplines that ABET accredits. You can discover which schools offer these programs by going to the ABET website, which includes a list of the schools that offer those programs.

7.12 Conclusion

One of the most important outcomes of this chapter is for you to understand the breadth and depth of the careers available to engineers. Since your career (see our chapter on career management) really begins with this course, it is important for you to really know all of these career possibilities in order to make more informed decisions. As you progress through your other courses on your way to your bachelor's degree and later on your graduate degrees or work experiences, you should review what you learned from the assignments from this chapter.

One of the most important outcomes of this chapter is your understanding of the incredible breadth and depth of the careers available to engineers.

We also reviewed five of the growth areas for engineering employment in the twenty-first century. Those new jobs will be in manufacturing, healthcare, infrastructure revitalization and enhancement, energy generation, and the environment. Finally we learned a great deal from reviewing the subdisciplines with the various engineering societies.

A very important area for engineering employment that we have not addressed is in the military or associated with the military. All of our military academies stress

IEEE societies http://www.ieee.org/societies_communities/societies/about_societies.html
Aerospace and Electronic Systems Society
Antennas and Propagation Society
Broadcast Technology Society
Circuits and Systems Society
Communications Society
Components, Packaging, and Manufacturing Technology Society
Computational Intelligence Society
Computer Society
Consumer Electronics Society
Control Systems Society
Dielectrics and Electrical Insulation Society
Education Society
Electron Devices Society
Electromagnetic Compatibility Society
Engineering in Medicine and Biology Society
Geoscience and Remote Sensing Society
Industrial Electronics Society
Industry Applications Society
Information Theory Society
Instrumentation and Measurement Society
Intelligent Transportation Systems Society
Magnetics Society
Microwave Theory and Techniques Society
Nuclear and Plasma Sciences Society
Oceanic Engineering Society
Photonics Society
Power Electronics Society
Power and Energy Society
Product Safety Engineering Society
Professional Communications Society
Reliability Society
Robotics and Automation Society
Signal Processing Society
Society on Social Implications of Technology
Solid-State Circuits Society
Systems, Man and Cybernetics Society
Ultrasonics, Ferroelectrics and Frequency Control Society
Vehicular Technology Society

Figure 7.7
Societies of the IEEE.

IEEE councils
http://www.ieee.org/societies_communities/societies/about_technical_councils.html

Biometrics Council
Council on Electronic Design Automation
Nanotechnology Council
Sensors Council
Council on Superconductivity
Systems Council
Technology Management Council

Figure 7.8
Councils of the IEEE.

Aerospace Engineering
Agricultural Engineering
Architectural Engineering
Bioengineering and Biomedical Engineering
Biological Engineering
Ceramic Engineering
Chemical, Biochemical, Biomolecular, Engineering
Civil Engineering
Construction Engineering Electrical, Computer, Engineering
Engineering, General Engineering, Engineering Physics, Engineering Science, Engineering
Engineering Management Engineering
Engineering Mechanics Engineering
Environmental Engineering
Fire Protection Engineering
Geological Engineering
Industrial Engineering
Manufacturing Engineering
Materials (1), Metallurgical (2) Engineering
Mechanical Engineering
Mining Engineering
Naval Architectural, Marine Engineering, Engineering
Nuclear, Radiological, Engineering
Ocean Engineering
Petroleum Engineering
Software Engineering
Surveying Engineering
Systems Engineering

Figure 7.9
A list of the engineering programs that ABET accredits.

degrees in engineering as very important for future officers. In addition to training officers in the field of engineering, our military also employs many civilian engineers to work closely with the military engineers.

END OF CHAPTER REVIEW QUIZ

Select the most appropriate answer to the following statements.

1. As a profession, engineering is
   a. A well-established and mature profession.
   b. A profession in decline in terms of employment opportunities.
   c. Just beginning to grow and to prosper.
   d. Entering a new and exciting phase of growth and transformation.
2. The best way of learning about different engineering disciplines is to
   a. Join and be active in one or more engineering societies.
   b. Search the Internet.
   c. Speak with faculty members.
   d. All of the above.

3. Of the top ten paying jobs available to undergraduates, engineering accounts for
   a. Three out of the ten.
   b. Six out of the ten.
   c. Eight out of the ten.
   d. Ten out of the ten.

4. Most of the jobs in manufacturing will be in
   a. Manufacturing engineering.
   b. Systems engineering.
   c. Mechanical engineering.
   d. Almost all fields of engineering.

5. Truly affordable and high-quality healthcare will only come through the field of
   a. Medicine.
   b. Government bureaucracy.
   c. Engineering working closely with the medical community.
   d. All of the above.

6. The engineering jobs within healthcare will be related to
   a. Biomedical engineering.
   b. Electrical engineering.
   c. Mechanical engineering.
   d. All of the above along with many more from other engineering disciplines.

7. Engineering jobs related to revitalizing and enhancing the infrastructure will be
   a. Civil engineering with support from a number of other engineering disciplines.
   b. Civil engineering.
   c. Mechanical engineering.
   d. Systems engineering.

8. The engineering discipline(s) that will be involved in obtaining energy independence as well as economic stability is(are)
   a. Systems engineering.
   b. Electrical engineering.
   c. Nuclear engineering.
   d. All of the above plus several others.

9. Producing the kind of environment we all want will only come through engineering! Which engineers will have the greatest impact on creating this environment?
   a. Environmental engineers
   b. Civil engineers
   c. Mechanical engineers
   d. All of the above

10. In an age when engineers will play a very central role, many of the jobs in the five drivers for engineering employment in the twenty-first century will actually be engineers, because
   a. Engineers will need to design and develop ways of automating more routine jobs.
   b. Engineers will need to interact with the general population much like physicians and lawyers today.
   c. Engineers will need to make sure we have the best possible environment.
   d. Of all of the above.

EXERCISES TO DEVELOP AND ENHANCE YOUR SKILL SET

Homework Problems

Note that the majority of the homework problems are actually found as Practice Problems scattered throughout the chapter.

Exercise 7.1 If you did not know which engineering major to pursue, did this chapter help you to decide which major or at least narrow down your search to a few majors? If you have narrowed down your search, what are the pros and cons of each major? What process are you going to use to select a major?

Exercise 7.2 Modify your career plan to include what you have learned in this chapter. Identify and explain the modifications you have made.

Exercise 7.3 What advice do you have for those who may wish to go into engineering? What advice do you have for those who are majoring in engineering and wish to select a major?

CHAPTER

The Skill to Apply the Systems Approach to Engineering to Enhance Your Career

Plans are only good intentions unless they immediately degenerate into hard work.

—Peter Drucker

An important part of this book is the concept that following a structured approach produces better results more quickly and with less effort. Such an approach results in higher accuracy relative to solutions and designs. It results in better tracking throughout the process. This approach is called the systems approach, although it can also have other names. Briefly, we can break the approach into the following: understand and refine the problem to be solve; develop different strategies to solve the problem and then evaluate which approach or approaches should be followed; and finally, the results are evaluated and a decision is made as to whether or not the results arc acceptable. If they are not acceptable, a decision must be made whether or not the problem statement needs to be changed and/or the problem-solving strategy needs to be changed. Again, the process is then followed as before until acceptable results are obtained or a decision is made to not pursue a solution.

8.1 Introduction

Let us begin by telling you that we are in no way experts in systems theory. We do know that having learned enough about this subject and having worked with it long enough (over thirty years), we realize that it is one of the most powerful tools an engineer can have. In fact, everyone could benefit from applying the basic elements of the systems approach to almost every aspect of their lives. In this chapter, we introduce these elements and help you develop the ablility to use those elements to help you in formulating your tasks or projects and to allow you to achieve your desired results with a minimum of effort and in a reasonable amount of time.

LEARNING OBJECTIVES

By using the information and exercises in this chapter you will be able to:

- Understand the benefits of using a structured process for almost everything you do in life.
- Know the appropriate balance between planning, organizing, problem solving, and making it all work together.
- Understand the importance of defining what you need to do.
- Develop alternatives to problem solving.
- Define and develop constraints.
- Understand brainstorming as applied to the systems approach.
- Determine how well you have met all the system requirements and whether you have come up with a successful solution.
- Understand the balance between planning and execution.
- Understand the importance of documentation.
- Understand the importance of communication.

Everyone could benefit from applying the basic elements of the systems approach to almost every aspect of their lives!

An example of how this approach works is to consider how my oldest daughter was able to accomplish what she saw as a very daunting task. Sometime during middle school, she was given an assignment to write a paper for an English class.

We simply used the systems approach to fist pick an overall topic and then break the entire theme into paragraphs. All she then had to do was just focus on writing one paragraph at a time. She turned it in and earned the highest grade she had ever earned for a paper she had to write for an English class.

Let us look at another task that clearly could never have been accomplished without using the systems approach. In the early 1960s, President Kennedy charged NASA with sending a man to the moon before the end of the decade.

None of the technology to accomplish this existed at the time. NASA started with the task of landing a man on the moon and returning that man safely to Earth. If NASA had started with an entire project, it would never have happened. But it did, and the project was so successful, that no one ever died with millions of passenger miles flown (three astronauts, Gus Grissom, Roger Chaffee, and Ed White did die in an unfortunate experiment that that went horribly wrong).

This program would never have been successful if the NASA engineers had not used the systems approach. Fortunately, they used the process of decomposing the problem into manageable subsystems and components. To give you an idea of the scope of the problem, the final solution was the Saturn V rocket. It had over 3 million parts, stood around thirty-six stories high, and was built by the lowest bidder!

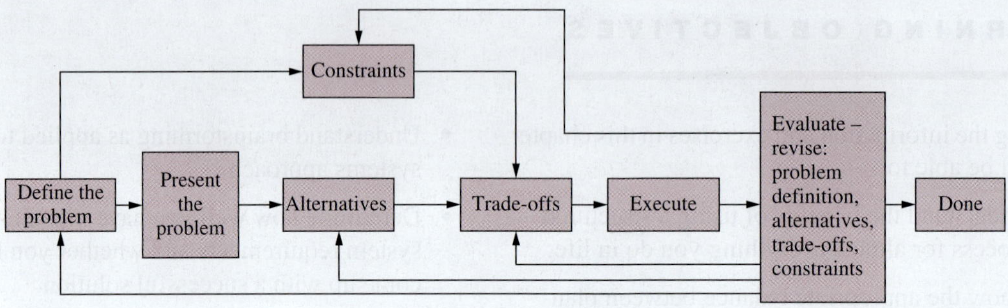

Figure 8.1
Flow diagram showing how the systems approach is applied.

As an aid to understanding the systems approach, a flow diagram of applying the systems approach is shown in Figure 8.1 This chapter covers the following elements:

1. Defining the problem or subsystem, including specifying the module interfaces (how the decomposed modules connect to each other and what are the values to be received and transferred)
2. Goals and objectives
3. Constraints
4. Alternatives
5. Strategies
6. Outcomes
7. Feedback
8. Documentation—also more thoroughly covered in our chapter on information capture
9. KCIDE
10. Top-down decomposition
11. Bottom-up realization
12. Defining the problem
13. Ready fire aim (Waterman and Peters)

8.2 Define the Problem or Task

"The need to clearly define the task or problem" is a theme that is carried out in other chapters with the goal that repetition is absolutely necessary in order to properly learn a skill. We need to realize that we must follow the "define"-and-"present," to whomever assigned the task or problem in the first place, process until we are satisfied we clearly understand the problem or task. We need to follow this process at every level to make sure nothing has changed, or if something has changed, that we have made the appropriate adjustments.

One of the characteristics of individuals, especially engineers, is the desire to complete a task. Whether it is painting a chair or writing a book or solving a homework problem, we want to just jump in and start doing it. How many times have you taken a test and just started working problems without first thinking about the problem, let alone the whole test. I guarantee that you will always do better if you will just stop for a minute and think about what you need to do to be most successful.

In all cases, a little effort at the beginning will save huge amounts of effort (not to mention frustration) at the end. You really need to develop an ability and a process to clearly define each and every task on which you will work. This is a really straightforward and a fairly simple thing to do, yet most of us routinely avoid it. First, it is important to know that clearly defining things normally requires an iteration process. You obtain the problem statement either verbally or better yet, in writing. You then need to convert this into a form you can understand. Using your own words and interpretation, ask if this is a clear understanding of the original problem statement. Hopefully, you will have written this down.

Let us start with some simple examples.

EXAMPLE 8.2.1

The problem statement is "Pick me up at the train station."

Clearly there is a lot of information missing. The most obvious is that we need to know, when and where. In addition, there are secondary pieces of information that could be important like, are you alone or in a group? If in a group, how many will I need to pick up? Will you have luggage? Where do you want me to take you?

Many times, we can realize some of the answers to these questions by knowing the context in which the problem statement is given. An example would be whether the train station is a passenger train station or a commuter train station. Also, the where may already be known if the individual is coming to stay with you.

Let us now look at your first work experience as either a new engineer or co-op student. Most of the time your supervisor will give you an assignment. Often that supervisor will have quickly come up with the assignment without giving it enough thought to clearly define the problem. Now you must work with the supervisor to clearly define the problem. Here you will find that at the initial meeting you will only be able to push this just so far. At some point, you will know that you will have to thank the supervisor and go back to your desk.

Once you are at your desk, what do you do next? You probably have a problem statement that falls into one of two categories:

1. The problem statement is fairly well defined and you can proceed to more fully developing the problem statement. We expand on this later.
2. You still do not have a clear problem statement. Thus, you still need to know more before you can proceed.

Let us look at the second issue first. You should look at this as a real opportunity to show how good you are! I would suggest that you write down exactly what you understand about the current problem statement. Now write down what you think you need to know before going further. You could stop here and submit what you have to your supervisor. I do not recommend this. My experience has shown that you really need to development some ideas first, develop these, and then submit your revised problem statement to the supervisor. You may then be surprised when the supervisor either accepts your ideas as reflecting his own and that this is what the supervisor meant all along, or you may be complimented on taking the initiative. In which case, the supervisor will give you a green light or give you a new problem statement, which will be much more specific than the original one. In either case, you will be a winner.

EXAMPLE 8.2.2 Expand the problem statement from Example 8.2.1.

In all cases, I would not stop here. First, continue to work on the problem consistent with the problem statement as it now stands, then take what you know and expand the problem statement to become more of a work plan. Once you have done this, run it past your supervisor just to make sure you are both of the same understanding of what needs to be done.

Original statement, "Pick me up at the train station."
First iteration (your questions that need answering)
Second iteration (your restatement of the problem statement)
Fully developed problem statement (includes any response, if any, to the restatement of the problem)

Another important consideration of this part of the proces is that if this task or subsystem interfaces with other tasks or subsystems, we need to be extremely clear about all the interfaces between this task or subsystem and everything that it interacts with. In other words, we must specify the module interfaces (how the decomposed modules connect to each other and "what are the values to be received and transferred") and make sure that we adhere to those requirements.

8.3 Decompose the Problem or Task

Once the problem is clearly defined, you need to decompose the problem into simple, well-defined modules. We do have a problem at this point. This process can itself be executed using the systems approach, and for more complicated problems, the approach we are defining must be followed.

8.4 Define the Elements of the Solution Process

Now that the problem is decomposed into more manageable problem blocks (or modules), we can proceed to develop solutions for those blocks. The process we follow is to now define the elements of the solution process. The elements of the solution process are covered in Sections 8.5 to 8.9.

8.5 Goals

Understanding what a goal is, is fairly simple. It is what we want to do or to which we strive. But, in using the systems approach, a goal is much more. We have goals relative to the overall system and then as we break a complicated problem into smaller units those smaller units each have goals, which were not goals in the level above it.

This concept will become clear as we progress into the chapter and as you study the rest of the book. We adopt the following definition:

A goal is that which you wish to accomplish and/or achieve.

A good example of my goal is to solve a particular problem. Another example might be I wish to write a book.

This now brings us to the question, How do you determine a goal? This also brings us full circle to the systems approach. Let us return to the example of solving a particular problem. At first glance, this seems like a simple enough goal. However, where did it come from? Clearly, as a student we have a goal to graduate. And to graduate, we need to pass classes. In order to pass classes, we must study the book and work problems. You can now see that the goal of solving the problem actually came about from the much larger goal of earning a degree. This is the systems approach, you take a much larger problem and break it into smaller and smaller units until you have a fairly simple unit that you can readily solve in a reasonable period of time and with reasonable effort. Clearly, you can see how this will eventually allow you to earn your degree.

Goals can also actually be an outcome of using the systems approach at a higher level. Another example is I would like to go from Philadelphia to Washington, DC. That is a goal, but the method of transportation has not been determined. We chose Philadelphia because you can easily reach Washington, DC, by car, plane, and train. Each of these has their own merits, and depending on your constraints and personal preferences, may be the method you choose to use. By the way, you will learn that we call these various modes of transportation, alternatives. You may be surprised to learn that the train is both the most economical and most convenient mode to use. We will develop this example later in this chapter.

EXAMPLE 8.5.1

Plan a trip from Philadelphia to Washington, DC.

Continuing with this example, in the process of evaluating these alternatives, the mode of transportation becomes a goal for that module. One of the goals for the module might be to travel from Philadelphia to Washington, DC, by car.

Let us look at the example of solving a problem. An important part of the alternatives would be which specific technique should be used to solve the problem. Then evaluating the various techniques becomes necessary to determine the best technique.

In that evaluation process, we would pick each specific technique and then solving the problem using that technique becomes a goal.

One last important point relative to goals—select a goal that is something over which you have control. For example, a good goal for a course is to learn as much as you can about the material presented in the course. A bad goal would be to earn an "A" in the class. Clearly the instructor controls the grades you earn, so grades are good metrics but not good goals.

8.6 Constraints

Constraints are extremely important to this process. Constraints are those things that limit the use of problem-solving techniques and processes and limit the design techniques and processes. In addition, they limit or place boundaries on the outcomes as well.

What are the obvious constraints for solving problems? I do not know about you, but one of my first constraints would be time. If a single problem (which in the course is not a major assignment) would require 100 hours of my time to solve, I definitely would not work the problem. The result might be for me to set up the problem solution process and give that to the professor and go on to the next problem. Another constraint for this might be the solution technique. For example, we might be asked to solve an electrical circuit problem using nodal analysis. This constraint eliminates mesh analysis from our list of alternatives we might use to solve the problem. This also brings us to the issue of how large the circuit is. If you have even just five nodes in the circuit, you cannot easily solve this by hand, so the alternatives might be to use a software package to solve your problem (MATLAB, for example) versus a very time-consuming set of hand calculations. In addition, circuit analysis packages such as PSpice could be used.

EXAMPLE 8.6.1

What are the obvious constraints for Example 8.5.1?

What are the obvious constraints for Example 8.5.1? The most obvious would be that we are constrained to consider only car, train, or airplane. This eliminates such other modes as walking, using horses, and taking a bus. Another might be financial, although in some cases this would not be a consideration. Time to destination might be a constraint. Time of travel, for example, you need to be in Washington by noon on a specific date. These would be the most obvious constraints.

8.7 Alternatives

As you may have gathered from the previous section, alternatives are the various methods of achieving your goals. Goals are to solve a problem and to come up with a particular design. Alternatives, are how do you solve the problems (which techniques should you use) and how do you achieve a specific design (again, which techniques and/or process should you use).

What are the alternatives for solving a problem? Some obvious alternatives we might have for all problems would be to solve by using hand techniques (this by the way is still used in a lot of courses), calculator techniques, or using a specific software package. Additional alternatives would be problem specific, such as the circuits example. Here you can use a combination of hand techniques and MATLAB or a specific circuit analysis package such as PSpice.

EXAMPLE 8.7.1

What are the alternatives for Example 8.5.1?

In the example of going from Philadelphia to Washington, DC, what are the alternatives? The obvious ones were actually given to us, car, train, or plane. You might think you now have your alternatives. However, you do have more when you break out each of the original alternatives and make them goals, which will then have their own alternatives. Let us look at using a car as the alternative for a moment. If our goal is to go from Philadelphia to Washington, DC, by car, alternatives might be use your personal car, borrow a car from a friend or a neighbor, or rent a car.

8.8 Evaluation and Trade-Offs

Most of the time we do not wish to execute each alternative (although when it is reasonable to do so, we could execute each one). Here we need to evaluate each alternative and pick the one we will use to achieve our goal. This is where the trade-off comes into play.

Why do we say trade-off when we are evaluating which alternative to use? This comes from the fact that there is no such thing as a perfect alternative. Generally, each alternative has its good points and its bad points. Since we will end up with an alternative, we know that the other alternatives will have desirable traits that the selected alternative does not have or where the specific trait is not as strong. As an example, we may choose a faster solution process that is not as accurate as a slower one. Let us look at our examples.

Given the problem in Example 8.5.1, what are the obvious trade-offs? In calculating a solution of a complicated system of equations by hand versus using the computer, we can attempt to identify the trade-offs. Historically, computer solutions tended to not be as accurate as carefully done hand calculations. Now, however, computer calculations can be much more accurate than the most careful hand calculations. They can be more accurate than those achieved by the most accurate calculators. So, the obvious trade-offs relative to hand versus computer calculations is that the hand calculations are not as accurate and are much, much slower to perform. In fact, there is a point at which, for a certain level of complexity, hand calculations can never produce the correct result.

Therefore, if you are willing to trade off accuracy and speed of calculation you can choose to solve the problem by hand. Not all problems have trade-offs that are this easy. Sometimes the trade-offs lead to more confusion rather than clarity. When this happens, you need to seek out others who can help you with the evaluation process.

	Example 8.5.1		
	Costs	**Arrival time**	**Duration**
Personal car	$135	Any time	3 hours
Rental car	$150	Any time	3 hours
Train	$160	Every hour on the half-hour	1 hour and 15 minutes
Airplane	$250	Every hour on the hour	3 to 4 hours
The milage costs on the personal car comes from calculating $0.45 per mile times 300 miles			
The duration for the airplane takes into account the additional time required for security at the airport			

Figure 8.2
Table comparing the costs, arrival times, and duration of each alternative.

Obviously, in the process of making informed trade-offs, we need to know all of the relevant issues related to each of the alternatives. This means we must evaluate all of the strengths as well as the limitations of each alternative. This is best done by adhering to the relevant constraints.

EXAMPLE 8.8.1

Evaluate each alternative for Example 8.5.1 and then execute the appropriate set of trade-offs.

For example 8.5.1, evaluate each alternative in Figure 8.2 and then execute an appropriate set of trade-offs.

If your goal or constraint is to get to Washington in the minimum amount of time, then the train appears to be the best answer. However, upon further evaluation, this only is the best answer based on living near the train station. If you do not live downtown near the train station, you must take into account the time necessary to get to the train station and getting back from the train station. This then brings us to the 3 hours, or more time that the other modes have. If your goal is costs, the personal car comes out the winner unless you want to save wear and tear on your own car. In addition, unless you really like driving, driving in that kind of traffic requires someone with a lot more of a death wish than I possess. Additional trade-offs are that renting a car will take more time and effort than using your own car. We also have not addressed the issue of parking in Washington.

8.9 Outcomes and Feedback

Did you achieve the desired results? If not, then you need to ask yourself whether you need to make changes to alternatives, trade-offs, constraints, and so on. Once you have made those changes, determine whether the changes now lead to the desired result.

Actually this part of the process is a loop and does not stop until you are successful or when you feel that you will never be successful. Our experience is that most of us will eventually achieve a desired result, although it may not be the one we started with.

8.10 Brainstorming

This process is how you determine the various elements of the systems approach.

An obvious question, by now, is how we determine all of these things if they are not given at the beginning. We do this using a technique that will probably be familiar to many of you. Brainstorming is nothing more than developing many ideas in a "free" environment. The key here is "free." In brainstorming, we permit ideas to flow freely without constraint. We start with a goal and then have alternatives presented as to how we might reach the goal. The free part comes from not discussing or explaining the alternatives, that must happen at a later time. The process of presenting alternatives happens like a lot of this process, it may start off slowly and then peak and then taper off as we exhaust the list of possible alternatives.

> Brainstorming is a process where many ideas are developed in a "free" environment. Ideas must be allowed to flow freely without being subjected to evaluation or criticism!

This same process can be used on the development of constraints. Some of our constraints come from those who present our goals, but additional ones may come out of a brainstorming session.

8.11 Conclusion

> The systems approach is truly nothing more than a wonderfully organized approach to solving many of the problems we will face throughout our careers. Engineers who master this approach will significantly enhance their chances for success, reducing tension, and enhancing their "fun" level!

END OF CHAPTER REVIEW QUIZ

Select the most appropriate answer to the following statements.

1. The systems approach to engineering
   - **a.** Only works with engineers.
   - **b.** Can be used in almost every aspect of life.
   - **c.** Is primarily focused on solving engineering problems.
   - **d.** Is best used to correct engineering problems.
2. The most important part of the systems approach is
   - **a.** Defining the problem or activity.
   - **b.** Determining the alternatives.
   - **c.** Brainstorming.
   - **d.** Trade-offs.

3. Decomposing the problem or task means to
  a. Start at the bottom and work your way up.
  b. Start at the top and break the problem down into smaller elements.
  c. Simplify the overall problem or task.
  d. Use an expert to decompose the problem or task.

4. An example of an appropriate goal is
  a. Earning a grade of A in a class.
  b. Earning a substantial salary.
  c. Outcomes over which you have control.
  d. All of the above.

5. Alternatives are
  a. Different ways to define the problem or task.
  b. Different ways to accomplish a solution to the problem or task.
  c. Different outcomes.
  d. All of the above.

6. Brainstorming means that
  a. You develop alternatives without discussing them or justifying them.
  b. Each alternative should be discussed and justified at the time it is brought up.
  c. Each alternative needs to be voted on as soon it is brought up.
  d. All of the above.

7. Examples of normal constraints are
  a. The number of individuals who are available to work on a project or task.
  b. The number of hours each individual can assign to a problem or task.
  c. The funding available to accomplish the problem or task.
  d. All of the above.

8. Trade-offs
  a. Sometimes are necessary to attain your goals.
  b. Usually are not important.
  c. Should be done regardless of whether or not they are needed.
  d. Are all of the above.

9. The field of engineering to which the systems approach is important is
  a. Systems engineering.
  b. Civil engineering.
  c. Mechanical engineering.
  d. All engineering disciplines.

10. The systems approach is truly nothing more than a wonderfully organized approach to solving many of the problems we will face throughout our careers. Engineers who master this approach will significantly enhance their chances for
  a. Success.
  b. Reducing tension.
  c. Enhancing their "fun" level.
  d. All of the above.

EXERCISES TO DEVELOP AND ENHANCE YOUR SKILL SET

Exercise 8.1 Write a story using this approach.

Exercise 8.2 Solve a math problem using this approach.

Exercise 8.3 Solve a physics problem using this approach.

Exercise 8.4 Make a decision on which class to take using this approach.

Exercise 8.5 Do a time management problem using this approach.

Exercise 8.6 Solve one of your problems from one of your engineering science courses using this approach.

Exercise 8.7 Solve one of your problems from one of your engineering courses using this approach.

Reference

In Search of Excellence: Lessons from America's Best-Run Companies, Thomas J. Peters and Robert H. Waterman, Jr., HarperCollins Publishers, February 19, 2004.

System Analysis, Design, and Development: Concepts, Principles, and Practices, Charles S. Wasson, Wiley, December 23, 2005.

CHAPTER

9 Project Management and How to Use It as a Skill to Enhance Career Success

> *"Of all the things I've done, the most vital is coordinating the talents of those who work for us and pointing them towards a certain goal."*
>
> —Walt Disney

In this chapter, you will learn the elements of project management and how those elements can help you become a better engineer. In engineering, you are hired for what you can do and not for what you know. In addition, you will always work in groups and on projects. The more you know about project management the better you can perform your tasks. You will learn about the entire project management process. The most important elements of this process are planning; organizing; motivating; directing; controlling; and finally, monitoring, evaluating, and revising all the elements. You will also learn how to take advantage of opportunities to enhance project management skills and the impact of project management skills on career success.

Project management is a lot like managing a budget. Some people think that developing a budget is nothing more than identifying your expenses and then that becomes your budget. We guess we all would like to live in such a world. In actuality, you have a fixed income and you then must budget those resources, and this becomes your budget. In project management you basically do the same thing.

Another important element of project management is that it heavily uses the principles of the systems approach. We explain this later in more detail.

9.1 Introduction

Early on in our pursuit of helping students and working engineers enhance their careers we realized the importance of project management skills in this effort. In fact, we cannot think of a project that will not benefit from using elements of project management. We discuss significant examples of when elements of project management made a significant difference. One of these is the leadership program we developed for IEEE Student Branches. The leadership program was developed totally around the elements of project management.

LEARNING OBJECTIVES

By using the information and exercises in this chapter you will be able to:

- Understand why project management is important for career success
- Develop an understanding of the project management process
- Better understand the importance of planning
- Understand the management of resources
- Understand the importance of milestones and deadlines
- Understand the importance of motivating yourself and your staff
- Control costs and resources
- Monitor and revise elements of project management
- Develop opportunities to enhance project management skills

This program was central in helping student membership in the IEEE double in only 3 years. We also use these elements to enhance the learning experiences in graduate-level engineering courses. Finally, we applied these elements to our Career Planning program, discussed in the chapter on career management skills. As you will see, we also applied these elements to other structured approaches used for a variety of applications.

First, we need a working definition of what constitutes a project. Essentially a project has a beginning, a middle, and an end. A project can be as simple as solving a homework problem or as complicated as starting a brand new manufacturing facility. For students, your senior design project will probably be the most complicated project on which you will work. We learn more about developing and executing senior design projects later in this chapter. Examples of projects are

- Solving a specific homework problem.
- Developing a career plan.
- Developing a time management plan.
- Complete a senior project.
- Developing a plan for student organization programs for an academic year.

Examples of activities that are not projects include

- Maintaining a career plan.
- Implementing a process to solve homework problems on an ongoing basis.
- Implementing a time management plan.

We must emphasize that when we refer to project management we are looking at efficient use of those tools and that we abhor becoming overly fixated on those tools. When you properly use the elements of project management you will realize

- Significantly better results.
- Being able to complete the project in a shorter period of time.
- Completing the project with fewer resources.
- Having much lower frustration levels throughout the project.

9.2 Why Project Management Is Important to You

Our human nature is essentially programmed to, when given a task, immediately jump into completing that task with very little thought about planning how we might complete that task. Let us ask you how many times you have taken a test where you immediately begin solving the first problem and want desperately to complete it before beginning the second problem. If you are anything like the rest of us, that is how you approached problem solving throughout your academic career until now. However, a much better approach might be to first set up the problem and then determine how you intend to solve it. Then do the same for each of the other problems in the test. Next, you work each problem until you need to plug in numbers to obtain a solution. Finally, you go back and then plug in the relevant numbers and solve each problem. A last step would be to check each problem. It should be obvious that following this process, especially considering tests where partial credit is given, will lead to earning higher grades with less overall effort.

These same advantages are valid when applying project management techniques to any project you need to undertake. However, there are numerous examples where an inappropriate implementation of project management techniques has lead to failed projects.

Again, we wish to advise you to be careful of becoming fixated with the tools to the point where project management techniques get in the way of easily and efficiently completing your project.

Why is it important for students to consider using project management techniques?

As a student, applying project management techniques to homework problems, quizzes and tests, laboratory experiments, and senior design will generally result in:

- Higher grades
- More efficient use or your time
- Much less frustration and maybe more fun with your studies
- Greater overall learning
- Happier instructors
- Other students wanting to work with you

9.3 The Project Management Process

For the purpose of this chapter we define the elements of project management as

- Defining
- Communication
- Planning
- Organizing

- Motivating
- Directing
- Controlling
- Monitoring, evaluating, and revising

9.3.1 Defining

Projects begin with defining what the project is. This part of the process can be started by you or the one assigning the project to you. Time spent here will be well rewarded later once the project is underway. Many times those assigning or developing projects do not have a clear understanding of what needs be done, so you must make sure you have asked enough questions and gotten enough of the right kind of answers to have a thorough understanding of the project. Actually, the defining of the project may further evolve as the project is being completed.

You need to have an idea of the expected date when the project can be completed. It is important to understand this completion date must reasonably reflect the reality of how long the project will take. Often we want to complete a project by a date that is not realistic. For example, if your project is to mow the lawn and mowing that lawn takes 3 hours with the mower you have at hand, then expecting it to be mowed in 5 minutes will never work no matter what you do. The completion date will depend on many things including the resources you have to help you complete the project.

Another part of defining the project is identifying the stakeholders as well as others who can impact the project. Stakeholders are those who are associated with the project in an active manner. They will be part of completing the project. Others are those who are not necessarily actively involved in the project but are able to serve as advocates for what you are trying to accomplish or maybe who are just interested in what you are doing. Identifying all of these individuals is very important and as with defining the project, this group may expand or contract as the project is being completed. For convenience, we will call this group the "contact group." Some refer to the contact group as the audience.

> The contact group includes the stakeholders as well everyone else with whom you feel you need to communicate.

9.3.2 Communicating

We cannot emphasize enough the importance of communication in project management. You must determine how you will communicate with your contact group. We will look at how you can accomplish this adequately.

> The most important element in project management is in communication!

Adequate communication includes but is not necessarily limited to

- Face to face meetings.
- Group meetings.

- Memos.
- E-mail.
- Conference calls.
- Reports.

One of the best written forms of project management communication instruments is called the "one-page project manager (OPPM)." This is available for use from www.onepageprojectmanager.com for a small cost. It is definitely worth using.

It is most important to have a clear communications plan as an integral part of the overall project management plan. Although this may seem unusual, even if your contact group is only you, having a clear communication plan with yourself will reap benefits later on. This is especially true if you develop something patentable or copyrightable. I suggest that the OPPM can be used as a core part of the communications plan.

Why plan on face-to-face meetings? Although this can be spontaneous, it is advisable to do this when possible on a regular basis. Regular meetings (maybe weekly) should be planned for the entire team working on the project. You should limit the time spent at those meeting to at most one and one-half hours. Remember your task is to complete the project. These meetings should focus on problems being encountered where input from the rest of the group may help solve those problems. Status reports should only be necessary when there is a problem with meeting task completion targets.

Written reports will need to be updated on a regular basis. Charts such as Gantt charts are very easy to read and understand and will give everyone a snapshot of the status of the various project tasks.

We suggest that at least 10% of your time be devoted to communication. I would also suggest that if you exceed 25% of you time communicating, you are either the supervisor of several individuals working on the project or that you are spending too much time communicating.

9.3.3 Planning

As you may have already guessed, the activities we are discussing have a great deal of overlap. We have discussed planning as part of the defining process and also a part of the communications process. Here, we specifically look at planning as in how to plan the execution of our project. We expand on the documents we developed as part of defining the project and our communication plans and develop a detailed plan as to how we will complete the project. An important part of the plan will be developing a set of constraints. Completing the project will require

- People resources in terms of the number of individuals that are needed to complete the project and a realistic estimate of the number of people actually available. Hopefully the two numbers will be close to being equal.
- People hours resources in terms of the number of work hours that each of the individuals can bring to the project as well as an estimate of the total number of people hours the project will require. Again, we hope the estimate is not larger that what is actually available.

- A realistic estimate of the funds that completing the project will require and a determination if those funds are available.
- A clear understanding of all the other resources that will be required such as space and equipment.

An important part of this plan is the identification of significant project elements and a timeline for the completing of these elements or task. This will serve as a basis of a Gantt chart as mentioned in the communications plan.

We would also like to see, as part of the plan, a narrative of how you expect to complete the project and what your expected outcomes are. This narrative can be useful in your communications with others.

In the process of completing your plan, you need to share it with those who can give you valuable feedback. It is important to get as much help as possible in doing this so that all problems and shortcomings can be identified as early as feasable. Once you have a complete plan, you need to have it approved by the appropriate individuals.

9.3.4 Organizing

Although you can begin organizing the project team and activities before the plan is approved, once it is approved you need to organize everything as quickly as possible and begin working on the project. Notice that you actually begin working on the project with a well-developed plan, a clear definition of the project, and a well-developed communications plan. You need to assign the tasks to the appropriate individuals and make sure they fully understand what is expected of them. One approach here might be to seek input from your team as to on what tasks might individuals like to work. The assignment of responsibility is one of the most important parts of managing the project. It is much like leading an orchestra. You need to make sure the assignments are in line with the talents of each of those who are responsible. Sometimes this is not possible, so a good leader needs to identify mismatches as quickly as possible and to make, tactfully, adjustments.

9.3.5 Motivating

It is very important to see that motivating the team is identified as one of the eight elements of project management. We are all human beings. Engineers just like everyone else do much better if they are enjoying what they are doing and see their tasks as being an important part of an important project. Just think of an engineer developing a part not knowing it was to be used in NASA's lunar landing program. How much better would it be if that engineer actually knew what an important role her part would play in safely sending people to the moon!

The first step in motivating others is to be excited yourself. Remember that it is very difficult to fool people; you sincerely need to be excited about what you are doing. Clearly letting others know how you feel about the project and the role that they are playing in the project will be very important. A major part of motivating others is in the quality of the communications (again we see the importance of effective communications) you have with them. Please realize a very valuable part of effective communications is in *listening* as well as in speaking. You can learn a lot from your team and what motivates them from listening to them.

Actually, motivating others is not as difficult as it first seems. Motivating others does not mean you need to be a superstar motivator like some of the people you see on television. Engineers are highly motivated to begin with. You can use that fact effectively by making sure your team is fully engaged in the project and their contributions are acknowledged appropriately. Seek their input frequently about the status of the project and make them feel they can make suggestions as to how the project can be improved.

9.3.6 Directing

> *"If you always blame others for your mistakes, you will never improve."*
>
> (Joy Gumz)

You are the one responsible for the project success. If it fails you will be the one who gets the blame. It is important for a good leader to make sure the members of the team receive the credit when the project is a success.

Directing the activities of the project, managing the project, and motivating the team, require leadership skills. Much of what is necessary in motivating the team was covered in Section 9.3.5. When projects are small enough that will allow you to be an active member of the team, you can lead by doing, which can be most effective. If the project is too large for this to happen, you now must assume the role of the director of an orchestra. Like the director, your main task is to make sure everyone is an effective member of the team. In addition, you must resist the tendency to put too much pressure on members of that team in the mistaken notion that pressure and tension increases productivity of creative and talented individuals. In fact, this causes a completely opposite reaction. We have found that calm and levelheadedness creates a much more productive environment!

Good directors and good leaders exhibit confidence and remain calm even when things seem to be falling apart. This brings us to the point of making sure your team stays focused, enervated, sharp, and in good spirits. Overworking your team, over stressing your team, and creating internal competition within your team are all counterproductive and must be avoided. Make sure deadlines are being met. If they are not, determine why. If it turns out that enough resources were not made available, make sure they become available or change the timeline. If it turns out that the task was not well understood and needs changing, then change it. If the wrong individual was responsible for the task, tactfully change the responsibility.

Remember that almost any problem has a reasonable and straightforward solution; all you need to do is find it. One of the things we suggest you follow when trying to solve problems is to resist the urge to determine whether you can solve the problem, something we all tend to do. Rather assume you have already solved the problem and then ask yourself how you solved the problem. You will be surprised how effective this perspective can be.

Just following your project management plan and following your communications plan will help you to provide appropriate direction and leadership. Again, listening to your team will often let you know what you should be doing.

9.3.7 Controlling

We have already mentioned some of the most important things to control while the project is being completed. Some of the most important of these are

- Motivating your team
- Making sure communications are maintained at a high level and are simple, straight forward, and clear
- Minimizing tension and competition between team members
- Making sure deadlines are reasonable and are being met
- Making sure resources are being expended in a reasonable and appropriate manner

9.3.8 Monitoring, Evaluating, and Revising

Monitoring your project progress is extremely important and again, communications is a key element in this part of project management. Work closely with your team to track the progress of each task and quickly deal with problem areas. If a task is not in a critical stream, then you have more flexibility in resolving it and we will address that soon. If the task is in a critical stream and threatens the completion of the project on time and within budget, then the problem needs to be dealt with immediately. Quickly organize those members of your team who can help you resolve the problem and get your project back on schedule. If the problem is an issue of resources, you need to determine if you can easily obtain the additional resources. If you cannot, then you may have to scale back expenditures in other areas.

If the problem is technical, then it must be resolved or its impact on the rest of the project needs to be minimized. Rarely will you find that technical problems doom a project. Remember, you have already solved the problem, you just need to determine how you did it!

Once you have resolved the issue, you need to reevaluate the project plan and make any necessary modifications and communicate what you have done with everyone else. This you need to do regardless of the nature of the problem.

If the problem is in a noncritical path, then you have more time and flexibility in what you need to do. If the problem is resource related, then you need to determine if extending completion dates will help. If not, can that task be revised to keep its resource requirement reasonable? If it cannot, then should the task actually be eliminated? If it can, then eliminate it.

If the problem is technical, then can revising the schedule help? If it cannot be, then can the task be eliminated? If so, eliminate it.

Evaluating your progress and resource commitment is a continuous process. Your various plans will help in evaluating everything. Evaluating your progress and resource commitment will help you stay on top of problems and will alert you to potential problems and issues that you need to resolve.

Monitoring and evaluating will serve as a basis for any revisions in your various plans. Once the project is complete your plans and communications will help you close down your project and report the results.

9.4 Opportunities to Enhance Project Management Skills

Let us summarize the skills that are required for good project management. Although there are other skills to help you to manage projects successfully, the following are the skills we consider to be the most important

Significant skills needed for successful project management:

- Communication skills
- Planning skills
- Organizational skills
- Motivational skills
- Leadership skills
- Team building skills

Many of the chapters in this book deal with ways to enhance these skills. Here, we simply suggest working within you student organizations to enhance your skills while in school and doing the same once you graduate. Working with your professional organizations allows you to work on all these skill without any significant risk. It also will result in an expanded network, which is also necessary for a successful career.

9.5 Impact of Project Management Skills on Career Success

Acquiring the skills necessary for good project management will in themselves lead to a more successful career even if you do not ever use them for a major project. Clearly having these skills when faced with a major project will significantly enhance your chances for success and will certainly lead to a more successful career and possibly lead to significant management opportunities (if that is in your career plan).

A close look at these skills reveals that they are also the skills necessary for a successful professional (engineer in your case) career in general. An additional by-product of developing and enhancing these skills will be an increased satisfaction with your work experience and even to your having more fun at work!

9.6 Conclusion

Although there is a lot more you can learn about project management from reading more books and even taking courses and pursuing a degree in project management, we have given you an excellent start and more than enough information to allow you to take on small- and medium-size projects.

One thing we did not cover is the variety of software packages available to support project management. We spend a minute here to discuss software. First, if you have never used project management software, it is not advisable to start using it when you are in the process of managing a project. It is much better to learn how to use the software well before you start a project, otherwise you may doom your project right from the start.

Some companies support the use of specific software packages for managing their projects. If your company does, you should thoroughly learn the software well before you begin your first project.

As a review and summary, the elements of project management are as follows.

The project and all the elements related to it need to be fully **defined**. At the same time, your "contact group" (stakeholders and everyone else you feel you need to communicate with) must be **defined**. An important part of the stakeholders will be the team that will actually complete the project.

A **communications** plan needs to be developed and followed. At the same time you should begin the process of **communicating** with your "contact group."

A detailed project management **plan** needs to be completed and followed. The **plan** needs to include reasonable estimates of the resources that will be required. These include the number of people needed for the project, the number of hours each individual can expend on the project, the amount of money needed for successful completion of the project, and any other resources that will be needed.

Once the plan is approved, the team that will complete the project needs to be **organized** and the project needs to be started. Resources will need to be **organized** as well.

A key part of executing the plan and completing the project well lie in **self-motivation** and **motivation** of the members of the team.

During the execution of the project, the project manager will be **directing** the various activities of key tasks necessary for completion of the project.

During the execution of the project, the project manager will be responsible for **controlling** the allocation and expenditures of resources. The project manager will also be responsible for **controlling** deadlines for the various tasks.

Finally, the project manager is responsible for **monitoring** the various elements of the project and their progress, for **evaluating** the progress of the various elements of the project, and for **revising** the project plan when necessary, **revising** the allocation of resources, and **revising** completion dates for the various elements of the project.

END OF CHAPTER REVIEW QUIZ

Select the most appropriate answer to the following statements.

1. Engineers are primarily hired for
- **a.** What they know.
- **b.** What they can do.
- **c.** How well they can lead projects.
- **d.** How well they work in a team.

2. Elements of project management are important to
   a. Project engineers.
   b. Systems engineers.
   c. Civil engineers.
   d. All engineers.
3. Understanding project management is important because it will help you
   a. Better perform your engineering tasks and problem solving.
   b. Work in engineering teams;
   c. Understand "success" in these projects or tasks.
   d. Achieve all of the above.
4. The most important elements of the project management process are
   a. Defining, communicating, planning, organizing, motivating, directing, controlling, monitoring, evaluating, and revising.
   b. Planning, brainstorming, alternatives, trade-offs, and outcomes.
   c. Planning, communicating, team building, and goal establishment.
   d. Planning, organizing, motivating, directing, controlling, monitoring, evaluating, and revising.
5. Project management draws heavily from which engineering discipline?
   a. Manufacturing engineering
   b. Systems engineering
   c. Industrial engineering
   d. Mechanical engineering
6. The most important constraint that project management must face is
   a. Time.
   b. Available resources.
   c. Actual project assignment.
   d. All of the above.
7. As a student, applying the principles of project management will result in
   a. Higher grades.
   b. Less time spent outside of the classroom on each course.
   c. Less frustration.
   d. All of the above.
8. The most important element in the project management process is
   a. Planning.
   b. Organizing.
   c. Motivating.
   d. Directing.
9. What is the most important thing to keep in mind during the planning phase?
   a. Missing an important goal or constraint
   b. Not creating your team before the planning phase
   c. Spending too much time in planning (ready, fire, aim)
   d. Allocation of resources
10. The single most important element in the project management process is
   a. Planning.
   b. Organizing.
   c. Directing.
   d. Communicating.

EXERCISES TO DEVELOP AND ENHANCE YOUR SKILL SET

Exercise 9.1 Use the elements of project management to develop your program of study (courses you will need to complete before you can graduate) and the skills you will need to master before you graduate.

Exercise 9.2 Use the elements of project management to develop a program for enhancing your project management skills.

Exercise 9.3 Use the elements of project management to plan your senior design project and how to execute it.

Exercise 9.4 Use the elements of project management to develop a plan for managing one of your student organizations including a program of activities and managing those activities.

References

Absolute Beginner's Guide to Project Management, 2nd Edition, Greg Horine, Que, 2009.

A Guide to the Project Management Body of Knowledge (Pmbok Guide) by the Project Management Institute, 4th Edition, December 31, 2008.

Microsoft Project 2010 Step by Step (Step By Step (Microsoft)), Carl Chatfield and Timothy Johnson, Microsoft Press, June 21, 2010.

Project Management: A Systems Approach to Planning, Scheduling, and Controlling, Harold Kerzner, Wiley, 2009.

Project Management for DUMMIES, 3rd Edition, Stanley E. Portny, PMP, Wiley, 2010.

Sun Tzu: The Art of War for Managers, Gerald Michaelson, Adams Media Corp, 2001.

The Fast Forward MBA in Project Management (Portable Mba Series), Eric Verzuh, Wiley, 2008.

The One-Page Project Manager, Clark Campbell, Wiley, 2007.

CHAPTER

Team Building Skills

The Power of Team Building

"Teamwork is the fuel that helps common people to attain uncommon results."

—Jim Watson, PE, President, Watson Associates

TEAM BUILDING SCENARIO

"As the project lead of a cross-functional team, I was tasked to create a plan for a new consumer product that meets customer needs, hits cost and resource allocations, and launches on time.

Our team needed to meet an aggressive cost goal on the total solution (product, packaging, accessories, and other materials) to hit our target market price. This meant that we could not do things 'the same old way.' We had to get creative without sacrificing quality or the customer experience."

(Beth Moses,Electrical Engineer and Program Manager, Hewlett-Packard Company)

Beth Moses has outlined a team situation that is typical of many corporate projects. We will identify the principles involved in team building and then, in our discussion at the end of this chapter, learn how Beth applied many of these principles to this project.

10.1 Introduction

Engineering teams are frequently used to handle the complexity of designing large projects. This chapter is designed to prepare you to be a successful team member or leader.

Use opportunities to build team skills now.

As a student, your team building skills can be developed by your participation in laboratory teams, senior project design teams, and projects sponsored by engineering professional organizations. And, as you develop and enhance team building skills now, you will be better prepared to successfully handle team opportunities in coop or intern positions, and throughout your working career.

LEARNING OBJECTIVES

By using the information and exercises in this chapter you will be able to

- Understand and apply the principles of successful team building.
- Be prepared to participate in more challenging team projects.
- Observe and recognize personality traits in yourself and others.
- Match your personality traits to team activities.
- Recognize and use opportunities to develop team building skills for greater career success.

10.2 Team Building

Teamwork is a powerful tool for technical professionals, and teams often replace individuals as the primary unit of operation. This is especially true in the more innovative organizations in business, industry, government, research, and academia.

10.2.1 Team Definition

A team usually *is not* just

- A group of individuals who work in the same location.
- A group of individuals who work for the same person.
- A group of individuals who do the same type of work.

A successful team is more than a group of individuals.

A team *is*

- A group that shares a common assignment.
- A group that recognizes that it needs the contributions of all members.
- A group that is committed to achieving maximum results within constraints.

10.2.2 Characteristics of a Successful Team

A successful team is a unified group of individuals with special and unique talents who work together to achieve common objectives. Strong and effective teams see diversity of ideas as a good way to blend the various abilities and strengths of each individual to accomplish a greater result than just the sum of individual contributions.

Successful teams share several characteristics including:

- Clear objectives.
- Strong leadership and appropriate roles for each team member.
- Good communication, trust, and openness to new ideas.

- Cooperation and ability to deal with conflict.
- Ability to balance innovation, quality, and cost.

10.2.3 Team Advantages/Disadvantages

Although teams help corporations accomplish major goals, there are advantages and disadvantages when you participate in team activities. These can be summarized as follows:

Team Member Advantages

- Work on new and more challenging projects.
- Share resources and skills toward common goals.
- Participate in learning experiences that enhance personal careers.
- Practice interpersonal skills and develop leadership skills.
- Produce greater results from synergy in group discussions.

Team Member Disadvantages

- Need to give up freedom to operate independently.
- Need to consider different ideas of other team members.
- May need to sacrifice individual goals to achieve team overall goals.
- Can be a waste of your time if poorly managed.

10.3 Team Building Process

Establishing a successful team is more than just asking a few individuals to work together on a project. Team members should be carefully selected based on their skills and knowledge and their ability to work effectively with other members of the team.

As an engineering student, you have skills that are important in team building. One good example is your ability to solve problems by a logical, step-by-step process, and this skill can be applied to a team assignment. So, let's review the basic principles of team building and see how your skills can make this a successful venture.

10.3.1 Forming

The following activities are typically included in forming a team:

> - The team is formed by members meeting to become acquainted and select a leader
> - Personal information or agendas usually remain in the background at the start
> - Project objectives and specific goals are identified and established
> - Important ground rules for team operation are established
> - Roles are identified for each team member

Figure 10.1
Team forming activities.

As indicated in Figure 10.1, you can help form a more successful team by participating in defining important items associated with the project. The project objective and specific goals to accomplish that objective also need to be clarified at the beginning.

As a team member, you will be more productive when you understand your specific role, starting in the forming stage. We discuss team member roles in detail in Section 10.5. Roles during the forming stage are very important and set the tone for an effective team. You will help your team be successful when you:

- Contribute your ideas during the development of team objectives and ground rules.
- Volunteer to participate in team activities that use your strongest skills.
- Listen to others and encourage team members to work together and use teamwork skills.
- Are prepared to invest sufficient time and apply your skills to achieve objectives.

Equally important to identifying team objectives is the establishment of ground rules that set the foundation for discussion and help the team develop recommendations. Some samples of ground rules are shown in Figure 10.2.

Time invested in forming a team whose members work well together results in time saved during the project and will increase the level of harmony within the group. More importantly, a well-formed team arrives at better solutions to the project, and this can have a positive impact on personal careers for team members.

The forming stage should be accomplished at the team's first meeting. The next step, storming, takes significantly more time and often requires a series of meetings.

10.3.2 Storming

The term storming is a good indication that this part of the team process can be lively and perhaps even stressful for some members. The ability of the team to arrive at the best possible solution to the project is based on how well individuals participate in the storming stage.

Let's start with an overview of the storming stage, as shown in Figure 10.3.

TYPICAL GROUND RULES

Be on time for meetings
Let one member speak at a time
Share time use
Encourage new ideas from brainstorming
Listen to all opinions
Criticize *ideas* and *not people* who offer them
Offer solutions and not complaints
Keep team objectives in focus
Work toward an honest consensus
Support team decisions
Take responsibility for team assignments
Complete work assignments on schedule

Figure 10.2
Typical team ground rules.

- Individuals introduce their ideas and goals
- Members listen to other ideas and consider their value
- Members promote own ideas and question or challenge ideas of others
- Ground rules are used to minimize chaos

Figure 10.3
Storming activities.

When you join a team, you usually have some preconceived ideas about the project. Some of your ideas will be based on your skills and interest in specific parts of the project. Guess what? Other members will also have ideas about the project, and many of their ideas may be different from yours. Frequently, you'll be representing your area of work and your supervisor will suggest desired outcomes for the project solution.

Most team members arrive on the scene with an agenda and are motivated to influence the team decision to fulfill that agenda. In addition, you may discover that team members often have hidden agendas that reflect personal career goals.

Because team members have these different agendas, discussions will be more productive when members follow ground rules as project solutions are introduced. This is the time for each team member to propose and discuss new ideas.

You, and all other members, need to contribute so the team has the advantage of exploring many different ways to design the project. If all potential solutions are not discussed, the final decision may be less than desired. Therefore, it is important to allocate sufficient time so that many different ideas can be discussed.

The storming session is most productive through the synergy of hearing one idea and then developing a second related idea. This is called brainstorming, and is a proven way to expand options. Results of brainstorming are maximized when members do not criticize others for suggesting new ideas. Ground rules should encourage brainstorming. Time invested in introducing new ideas can produce better results, and the team should not be impatient when members want to expand the storming session.

Explore a multitude of options to arrive at the best solution.

The second important part of the storming session is for each member to defend his or her ideas and to offer documentation to support these ideas. As a member presents an idea, other members have three assignments—listen carefully, ask questions for clarification or additional information, and discuss the value of the idea. Though sometimes frustrating, this is the best method for a team to decide what should be included in their recommendations.

The storming session can be a challenging part of team building. When members are passionate about their ideas, they often see questions as criticism, and take comments to be rather personal. When this happens, the team leader needs to be a mentor, apply principles of conflict resolution, remind the team of the ground rules, and remove the personal aspect of discussions.

An important principle of the storming stage is that the team should take as much time as possible to arrive at a viable solution. One or more better solutions may be overlooked if many ideas are not offered and discussed before decisions are made. For team members who favor early decisions, this can appear to be taking too much time. However, the overall results are well worth the extra time investment in this session.

10.3.3 Conforming

To be efficient and meet reasonable time schedules for project development, the team needs to make the transition from discussing individual options to arriving at a team solution. This is called conforming because all members will likely have to make some concessions for the team to arrive at a single solution. Activities in this stage are listed in Figure 10.4. Recommendations will impact each member, so members should work hard to find the best team solution.

- ➢ Team leader takes control and moves team into a cooperative mode
- ➢ Hidden agendas may be disclosed to solicit support
- ➢ Individuals compromise to arrive at a team solution
- ➢ Team solution is clarified and members identify assignments

Figure 10.4
Conforming activities.

The responsibility of bringing the team to a single solution rests heavily on the team leader. Because individuals may have personal agendas, the leader needs to start this step carefully by focusing more on the concepts than the individuals who support them. This typically includes a give-and-take approach, and the process may now reveal some of the hidden agendas.

A major role of the team leader is to keep the team on track.

Eventually, the final solution is developed and all members need to agree to support team decisions. This is a time for effective communication to ensure all members understand the solution and their role in making it happen.

The team solution to the project should be put in writing and distributed to all members. This is the best way for you, and other team members, to have a clear picture of what the team decided and how each member will participate in implementing the solution.

And this brings us to the fourth and final step in team building—performing.

10.3.4 Performing

The major activities of the performing stage are shown in Figure 10.5.

- ➢ The team becomes committed to achieving goals and completing the project
- ➢ Members review assignments and start work in a cooperative spirit
- ➢ The leader reviews work progress and encourages completion
- ➢ Team members communicate results to appropriate management levels

Figure 10.5
Performing activities.

Attitude is important when team members reach the fourth step. If members who did not feel the team solution met their goals or the goals of their department have a negative attitude, they may not complete their team assignment well.

One way to minimize negative attitudes is for the team to develop supporting information to demonstrate why the team decision is best for all concerned. This will also help members communicate results to their departments and may prevent negative impacts on their careers.

The team leader should continue to coordinate the implementation of the project and remind members of their assignments, when needed. When each member meets the target dates for completion, everyone benefits.

Communication is an important key to success.

Communication during the implementation of the project is just as important as communication in team work sessions. The concepts discussed in the chapter on written communication can be applied to team reports and this will help to coordinate the project implementation. You can also add the project report to your records of achievement.

"Being a student leader with many different facets of my life was synonymous to being a judge at a talent show. I needed to indentify the potential and the talents possessed by fellow students who would otherwise not step forward to participate in various capacities. In no time I created a following and a culture of students with whom I studied and worked. This made it possible for me to delegate pressing issues that I could not personally attend to. For example, I formed my senior design team but made a colleague the leader of the team and our senior design team won the best senior design project of the year."

(Sedofia Gedzeh, Electrical Engineer)

You can prepare for team projects in your working career by taking advantage of opportunities to practice team building now. Some of these opportunities are laboratory classes and senior design projects. You can also practice team building when you participate in professional organization projects. These experiences can help you be a better team player in your working career.

10.4 Personality Traits

So far, we've discussed team building from the viewpoint of the team. Now, let's turn our attention to the most important part of any team—you and the other individuals involved.

Each individual is special and unique. We are different in our backgrounds, experiences, attitudes, and abilities. All of us have opportunities, challenges, disappointments, and moments of great success. Because we are unique, we respond to situations differently. And, personal differences can enhance our lives, because as we interact with others we learn new ideas and observe different personality traits.

Personality traits are not good or bad, they are just unique. The most important person for you to study and understand is you. Because you are unique, you can better relate to others if you first understand your own behavioral tendencies.

Each individual is unique and important.

Your knowledge of your strengths and limitations provides a foundation for achieving a professional attitude and work ethic. You can maximize your strengths, minimize your limitations, and learn to be effective in interpersonal relationships. And, when you understand your personality traits, you will be able to observe similar and different traits in others.

Behavioral patterns can be identified by professional assessment and interpretation. Perhaps you have been involved in this type of assessment and are aware of your personality traits. Our discussion in this chapter is intended to help you apply your most effective skills so you enjoy your involvement in a team setting and can be most productive.

10.4.1 Personality Traits Categories

Personality traits are typically identified by looking at four basic behavioral patterns associated with how you participate in group situations. There are many different titles for the four major personality traits. For simplicity, we will use the titles in Figure 10.6.

10.4.2 Personality Trait Strengths and Limitations

An overview of typical strengths of each of the four major traits sets the foundation for discussing how personality traits impact team building. For example, *drivers* typically are good at accepting challenges and taking responsibility of making things happen.

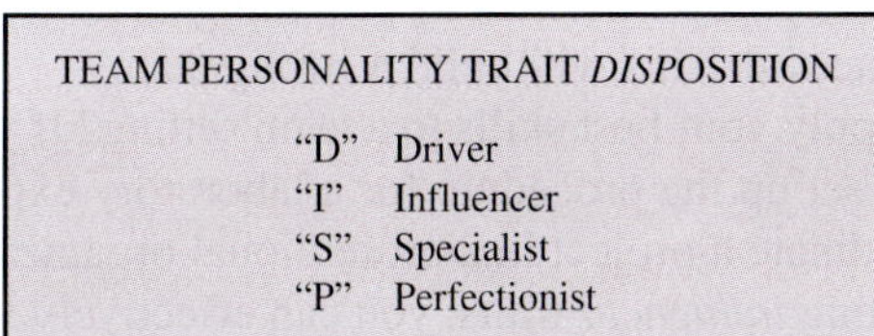

Figure 10.6
Personality trait categories.

They can visualize the future and take control of difficult situations by solving problems. Because they question present status, they often cause problems for people who resist change.

In contrast, *influencers* love to interact with other people and are good at expressing their ideas and making positive first impressions. They are helpful, are usually active in group settings, and can be quite entertaining. They bring enthusiasm to group settings and tend to talk more than listen. Because they do not worry about details, they are often perceived as being superficial.

As indicated by their title, *specialists* enjoy operating in a well-established world. They like to stay in a comfortable environment, concentrate on their part of a project, and use specialized skills. They are patient, loyal, sincere, and honest. They appreciate security and are good listeners. Because they do not like change, they are uncomfortable in thinking outside the box.

Perfectionists concentrate on details, follow standards, and like to be 100% accurate. They are diplomatic, enjoy being supervised by others, and accept decisions by others. They appreciate compliments for their good work but can be very critical of poor work performance by others.

Most people have a mixture of these traits, but usually one or two tend to stand out in personality evaluations. In each case, we can use our traits for successful results.

10.5 Personality Trait Application to Project Teams

A basic understanding of personality traits gives insight to better observe others and to learn the best methods of interacting with those around us. When we avoid trying to change others and work to understand and appreciate their traits, we help the team achieve much greater overall results.

Fortunately, we don't need a detailed analysis of other people to observe their strongest personality traits. Strong leaders often demonstrate *driver* traits. *Influencers* bring creativity and enthusiasm to a group. Team members who are *specialists* are excellent in completing work and will remain loyal to the project. *Perfectionists* are diplomatic and produce quality work.

Based on the results of our observations, we can learn to influence others or to increase cooperation within group activities by providing an atmosphere in which others are motivated and successful. This will build a strong team.

So, how can you apply your best skills to a team setting? If you have strong *driver* traits, you can offer to set up the procedure for a laboratory experiment, lead a senior design project, or coordinate a project in a professional organization.

Or, if you have strong *influencer* traits, you can effectively verbalize the results of a team project. Your strongest asset in a team activity is your enthusiasm and creativity. If properly channeled, your creativity will add to the success of a laboratory team or other group activities.

Influencers do not like details of a project, but are excellent supporters of the team. So, if you are an *influencer*, you will be good at communicating team needs and results and should play a major role in written and oral presentations that are an important part of all projects.

Each team needs one or more *specialists*. If you have strong specialist's traits, you are a good listener and can concentrate on the tasks to be completed. You will be most effective when the procedure is well defined and is not changed.

Specialists need to identify with the overall group, and are motivated by compliments from others. If you are a *specialist*, you are not usually aggressive so you may need help in getting started on the work. Your loyalty will encourage others to complete the task, and this helps to obtain results.

Because accuracy of results is important, the team needs at least one *perfectionist*. If you fit this personality type, you will enforce procedures, and this should help the team obtain accurate results. In group planning meetings, you tend to continually check for the correct procedure and accurate information. You review plans and instructions very carefully and prevent the group from making serious mistakes. As a *perfectionist*, you will be most effective if given opportunities to be precise.

Teams perform best when members apply different personality traits.

If all team members have the same personality traits, the team will be unproductive. A team whose members are all *drivers* will spend most of their time trying to elect a leader. *Influencers* will talk all day about a variety of topics but never establish a project plan. A team of *specialists* will wait forever for someone to make assignments. And, a team of *perfectionists* will just study the assignment and take no action.

So, the most productive teams are made up of members with a variety of personality traits. For best results, assignments should match the stronger traits of each member. This does not guarantee team harmony, but supports greater success in achieving project objectives.

10.6 The Value of Diversity

Teams are excellent opportunities for you to actively participate in decision-making activities and to realize a greater sense of accomplishment. Because teams consist of individuals, team success depends on how well individuals work together. Team results are directly related to how well team members include a diversity of ideas.

If your team consists of members from a diverse background with different ideas and skills, this can be used to great advantage when members view a variety of ideas as a way to expand options. Ground rules should encourage new and different ideas and viewpoints on topics. Members should give ample time and listen and respect all ideas, especially during brainstorming activities.

> ***"The true beauty of teamwork emerges when it offers a glimpse of creative solutions beyond the walls of your biases and preconceptions. For new teams, I've found the biggest stumbling block to achieving these insights is a failure to incorporate diversity of life experience in addition to skillset. Without this diversity, teams risk reinforcing lackluster ideas and producing mediocre outcomes. These failures often take teams by surprise, still unaware that they were working within their comfort zones the whole time."***
>
> **(Joshua Brandoff, *Mathematical Programmer and IT Specialist at the Museum of Mathematics*)**

Because individuals are different in attitudes, abilities, and experiences, they respond to team situations in different ways. The value of diversity is that different people have different ideas, and this expands options for the team to find the best solution to a project. Your contribution to a team will be valuable when you maximize your personality trait strengths and minimize your limitations.

A CONVERSATION WITH BETH MOSES

Beth, tell our readers how you used the principles of team building in your project introduced at the beginning of this chapter.

Answer:

"I will answer that question by indicating how we used the basic team building steps.

Forming:
Our team consisted of ten professionals ranging from software engineers to marketing managers. The members were experts with specialized skills to ensure a broad perspective across the whole customer experience. Each person brought diverse backgrounds and experiences.

As team lead, in our first meeting, I outlined the situation. . . the target customer and desired experience, project cost goal, timeline, the scope of what we could impact and fixed items that could not be changed.

Early in the project, it was important to spend time getting to know each other. We often did a 'round table' discussion that allowed each person sitting 'around the table' to share status as well as describe issues and concerns on the project. This gave each person a chance to talk and for us to better understand personality traits as well as team members' strengths.

Storming:
We identified drivers, influencers, specialists, and perfectionists on the team fairly quickly. This helped me as team leader to request tasks and activities of each team member that best matched his or her personality traits.

We spent time brainstorming and discussing ideas on how to exceed the customer's expectations, while best utilizing our team resources within the cost constraints. As team lead, I wrote the ideas on a flip chart to ensure that no idea was lost. We then performed a quick assessment of the ideas to determine impact to the customer experience, time frame required, resources needed, and cost to the solution.

This exercise involved everyone's diverse inputs and expertise to be sure that we evaluated the ideas from all angles. This step was crucial to include all members of the team, as no one person in our team had all of the answers.

Conforming/Performing:

In the end, our team determined several ways that we could deliver a better, higher-quality product by providing information in a different, more interactive way. To ensure that the customer was successful getting started with the product, we included a large, 'setup' poster with plenty of visuals to keep words and translations to a minimum. This solution helped to quickly deliver the necessary start up instructions, provide a great customer experience across many countries and cost less than previous approaches.

For more specific operating instructions, we enhanced on-screen instructions shown on the product display. By using visuals and illustrations right on the product, the customer was able to quickly access information and directions when needed in the operation of the product. Finally, we included a full written manual on the software disc. This way, the customer could read the step-by-step operating instructions on a computer and even print them out if desired.

The overall solution was a cost-effective approach and delivered a high-quality product to the consumer. In addition, we were able to leverage our resources in a new and different manner while still meeting timelines."

How would you summarize the results of your participation in this career opportunity?

Answer:

"Our team was able to find a creative solution that exceeded customer expectations while leveraging resources, achieving cost goals, and introducing to the market on time. By bringing together a diverse set of people who were focused on meeting the team's objectives, we created and delivered a successful product.

As team leader, I leveraged team members' expertise, but challenged each one to think differently to solve problems. The willingness of team members to partner with each other to create a new and better solution was the turning point.

As team lead, I also felt strongly that successes should be recognized and celebrated. Sometimes I would simply announce accomplishments during team meetings. Other times, I would bring candy or cookies to celebrate team successes (we loved this in kindergarten, why not in the workplace?). Bigger accomplishments deserved email notes to management. Finally, we celebrated major milestones with larger organization celebrations."

One final question—what advice would you share with students?

Answer:

- Early in the team formation, a team leader should provide clear team vision, objectives, time frame, and constraints. This helps align all team members to common goals and challenges.
- Bring together key team members with diverse backgrounds, expertise, and specialties.
- Spend time early forming team, getting to know the team members, understand strengths and personality traits.
- Listen to ideas and solicit inputs from those that are more quiet or shy—use the 'round table' approach where everyone in the team gets a turn to share status, successes, frustrations, and issues.
- Track, recognize, celebrate success (small and large).

(Beth Moses, Electrical Engineer and Program Manager, Hewlett-Packard Company)

10.7 Conclusion

The chapter on team building is included in this book because you will often be involved in team projects. The best way to prepare for future team projects after graduation is to use opportunities to develop and improve your teamwork skills now.

Some opportunities for you to build teamwork skills are included in your classes and laboratories. Others can be found by volunteering to be an active member of one or more professional organizations and participating in group activities.

We emphasized the value of developing nontechnical skills for team building because this will help you achieve greater career results. And, you will be more successful as a team member when you use engineering principles of logical thinking and organization.

We used engineering principles in this chapter to develop the logical steps in team building and in the application of personality traits to teamwork. So, apply a balance of technical and nontechnical skills in your future team opportunities and you will enjoy the benefits of working with others to complete successful projects.

END OF CHAPTER REVIEW QUIZ

Select the most appropriate answer to the following statements.

1. A team is

a. A group of people who work in the same department and have common interests.
b. Several engineers who work for the same person.
c. A group selected to develop a project or solve a problem.
d. All of the above.

2. A typical characteristic of a successful team is
   a. The selection of a strong leader.
   b. Good communication within the team and to stakeholders.
   c. The ability to balance many options to find a solution to a project.
   d. All of the above.
3. The most important advantage of being on a team is
   a. To get promotions by achieving the goals of your department.
   b. You can be a leader and tell others what to do.
   c. You can work on more challenging and rewarding projects.
   d. You don't have to spend as much time on routine activities in your job.
4. The biggest disadvantage of working on a team is
   a. You may need to give up some of your goals to reach a team solution.
   b. You may have to work with people you don't like.
   c. You might have to put in extra hours to keep up in your other work.
   d. You have to take notes and give a report to your supervisor.
5. You can determine your personality traits by
   a. Asking other people to tell you what they like about you.
   b. Using evaluation tools and by thinking about what you usually do best.
   c. Asking your supervisor why you are given various tasks to do.
   d. Doing all of the above.
6. The most effective skills of a person who has strong director traits are
   a. Leadership and making things happen.
   b. Thinking before deciding what to do.
   c. The ability to relax and not worry about what might happen.
   d. Getting involved in details so results are perfect.
7. People with strong influencer traits are not good at:
   a. Interacting with other people.
   b. Making a positive first impression because they talk too much.
   c. Being patient with people who like details.
   d. Showing enthusiasm for a project.
8. Specialists are important on a team because they
   a. Initiate new ideas.
   b. Are good at adapting to change.
   c. Can handle several tasks at a time.
   d. Are patient, sincere, and very honest.
9. People with strong perfectionist traits are not good at
   a. Delegating work to other team members.
   b. Concentrating on details or standards.
   c. Accepting decisions by others.
   d. Telling others when their work is not up to standard.
10. A successful team should consist of
    a. Several members with driver traits so the work will get done.
    b. Members with different personality traits to take advantage of diversity.
    c. Many members with perfectionist traits for greatest accuracy of results.
    d. All members who have strong influencer traits to have good communication.

EXERCISES TO DEVELOP AND ENHANCE YOUR SKILL SET

Exercise 10.1 Project team exercise

Option 1: Your instructor will discuss details and assign a team project exercise based on Chapter resource 10.1, Envirocar Project Team Scenario.

Option 2: Ask members of your study group to use Chapter resource 10.1, Envirocar Project Team Scenario, and practice team building skills.

Exercise 10.2 Project team exercise report

Project Team Report

If your instructor selects Option 1 of Exercise 10.1, review the results of your project team exercise and complete your project team report using the following format and outline headings.

Envirocar Project Team Report

Name ____(Your Name)____

Team identification:

Team member names:

Your assigned scenario:

Team solution:

Major reasons for selection of team solution:

In what team activities would you be most effective? Give reasons for your choices.

How did you participate in each of the team building steps?

Forming:

Storming:

Conforming:

What would be your most effective contributions to this project during the performing stage?

Exercise 10.3 Personality trait self-evaluation

Your instructor may assign this exercise or you can elect to complete it on your own. Use the self-evaluation form shown in Chapter resource 10.2, to determine your "*disp*osition" scores relating to a project team scenario.

Exercise 10.4 Project team report including application of personality trait evaluation results

Use your personality trait self-evaluation results from Exercise 10.3 to consider how your traits were applied in the project team Exercise 10.1, and complete your project team report using the following format:

Envirocar Project Team Report

Name ____________(Your Name)____________

Team solution:

Major reasons for selection of team solution:

Numerical subtotals of personality traits self-evaluation exercise:

Driver

Influencer

Specialist

Perfectionist

Based on your personality trait self-evaluation and skills, in what team activities would you be most effective? Give reasons for your choices.

How did you participate in each of the team building steps?

Forming:

Storming:

Conforming:

What would be your most effective contributions to this project during the performing stage?

END OF CHAPTER RESOURCES

Chapter resource 10.1

Envirocar Project Team Scenario

Contest overview: Your team goal is to design a race car that will achieve the highest total points in a 12-hour race. The car is to be constructed by students using school facilities and entered in a contest at a nearby Honda test track in which students from seven other universities will compete.

Contest details:

1. The race will be held in six months at a Honda Test Track from 10 AM to 10 PM.
2. Teams will be limited to eight people, including drivers.
3. Teams will be limited to a maximum of 20 gallons of regular unleaded gasoline.
4. Entries will be judged on a point system:
    - 500 points to complete the 12-hour contest.
    - Bonus of 1 point/mile completed.
    - Bonus points for lowest total car equipment, construction, and fuel cost:
        - 300 points for lowest cost of all contest cars.
        - 200 points for second lowest cost of all contest cars.
        - 100 points for third lowest cost of all contest cars.
    - Bonus points for environmental impacts:
        - 500 points for zero use of gasoline.
        - 300 points for maximum use of 10 gallons of gasoline.
        - 100 points for maximum use of 15 gallons of gasoline.

Race car designs are limited to the following:

1. Use an electric motor powered by solar photovoltaic cells on the car to charge batteries and interchange batteries as needed during pit stops.
2. Use an inexpensive internal combustion engine, aluminum body, 18-inch wheels, and high-pressure tires to obtain high fuel efficiency.
3. Use a new, experimental hybrid engine that is powered by gasoline and an electric motor powered by lead-acid batteries.

Team Building Exercise: Review assigned scenario details, prepare an outline of how to promote this scenario, meet with team, and follow the assignment steps.

Envirocar Scenario 1—Solar Car

Design details:

- Use photovoltaic cells located on the top of the car to charge nickel-metal hydride batteries.
- Use proven direct-current motor and simple speed control.
- Build car using fiberglass to reduce weight and steel for strength.
- Select body shape for maximum photovoltaic exposure and minimum air resistance.

Contest plans:

- Use small drivers to reduce weight.
- Change batteries as needed during pit stops.
- Provide battery charger and use existing track electrical outlets.

Potential advantages:

- Requires no use of gasoline so qualifies for 500 bonus points for zero environmental impact.
- Results in zero fuel cost during contest.
- Increases energy conservation and miles driven by low weight-to-power ratio.
- Provides opportunity to demonstrate engineering skills using futuristic systems and should enhance future job opportunities for team members.

Potential disadvantages:

- Estimated construction cost of $12,000 may be higher than other designs.
- Energy from photovoltaic cells will be reduced if weather is not sunny and will disappear after sunset a few hours before the end of the contest.
- The numerous electrical connections associated with photovoltaic cells increases construction time (600 hours) and could result in failures during the contest.
- Energy limitation of using only batteries after dark may limit ability to complete race.
- Top speed will be limited to 30 miles/hour.

Your role in this team exercise:

- Before meeting with your team,
  - Read outline of contest and details of this design scenario.
  - Outline your plan to discuss this design.
- During the forming process with your team,
  - Introduce yourself and record names of other team members.
  - Participate in selecting leader.
  - Offer ideas for ground rules.
- During the storming process with your team,
  - Briefly discuss this scenario and promote the advantages.
  - Be prepared to defend scenario when questions arise relating to potential disadvantages.
  - Listen to other scenario discussions and ask questions about potential disadvantages.
- During the conforming process with your team,
  - Compromise as needed to help team select final design.
  - Record advantages of final design so you can discuss with your team advisor.
- Complete the performing process.
  - Prepare a written report as assigned.

Envirocar Scenario 2—Aluminum Car

Design details:

- Use aluminum for car structure to keep weight low.
- Use large lawn mower internal combustion engine for low initial cost.
- Use 18-inch wheels and high-pressure tires for increased fuel efficiency.
- Keep car size very small to reduce energy consumption.

Contest plans:

- Use 20-gallon fuel tank to reduce number of pit stops.
- Select two small drivers and rotate to minimize fatigue.
- Limit top speed to 40 miles/hour to conserve fuel.

Potential advantages:

- High efficiency should support completion of contest and 500 points.
- Obtaining engine efficiency of over 40 miles/gallon may result in 100 point bonus.
- Total construction cost of $5,000 by using inexpensive engine and materials should increase cost bonus potential.
- Simple design should reduce construction time to 400 hours.
- Reduced pit stops should increase miles driven and bonus points.

Potential disadvantages:

- Making strong aluminum welds requires special skills.
- Most lawn mower engines are not designed to run at high speeds for 12 hours.
- Aluminum is more likely to be damaged if contact is made with cars or the track.
- Top speed of 40 miles/hour may limit bonus points for total miles driven.

Your role in this team exercise:

- Before meeting with your team,
  - Read outline of contest and details of this design scenario.
  - Outline your plan to discuss this design.
- During the forming process with your team,
  - Introduce yourself and record names of other team members.
  - Participate in selecting leader.
  - Offer ideas for ground rules.
- During the storming process with your team,
  - Briefly discuss this scenario and promote the advantages.
  - Be prepared to defend scenario when questions arise relating to potential disadvantages.
  - Listen to other scenario discussions and ask questions about potential disadvantages.
- During the conforming process with your team,
  - Compromise as needed to help team select final design.
  - Record advantages of final design so you can discuss with your team advisor.
- Complete the performing process.
  - Prepare a written report as assigned.

Envirocar Scenario 3—Hybrid Car

Design details:

- Use new experimental hybrid Honda system with gasoline engine and electric motor.
- Use standard lead-acid batteries to power electric motor.
- Design car with low air resistance for added fuel efficiency.
- Use automatic transmission to reduce driver fatigue.

Contest plans:

- Use 10-gallon fuel tank to reduce weight.
- Plan to run at top speeds of 50 miles/hour.
- Change batteries during pit stops.

Potential advantages:

- High average speeds should result in a large bonus for miles driven.
- Use of the hybrid engine will result in 300 bonus points.
- Use of Honda system builds relationship with company and may lead to job opportunities after graduation.
- University will receive goodwill by supporting new technology.

Potential disadvantages:

- Adapting new technology may result in problems during construction.
- The hybrid system will not be available until 2 months before contest—this may complicate construction estimated to be 300 hours for chassis and 200 for hybrid system.
- Untested new technology may limit miles driven and the ability to finish the competition.
- The high cost ($10,000) of this technology may minimize opportunity for cost bonus points.
- The frame to hold the engine, electric motor, and battery system needs to be very sturdy and may increase construction time.

Your role in this team exercise:

- Before meeting with your team,
  - Read outline of contest and details of this design scenario.
  - Outline your plan to discuss this design.
- During the forming process with your team,
  - Introduce yourself and record names of other team members.
  - Participate in selecting leader.
  - Offer ideas for ground rules.
- During the storming process with your team,
  - Briefly discuss this scenario and promote the advantages.
  - Be prepared to defend scenario when questions arise relating to potential disadvantages.
  - Listen to other scenario discussions and ask questions about potential disadvantages.
- During the conforming process with your team,
  - Compromise as needed to help team select final design.
  - Record advantages of final design so you can discuss with your team advisor.
- Complete the performing process.
  - Prepare a written report as assigned.

Chapter resource 10.2

Personality Trait Self-Evaluation

*Determining Your "**DISP**osition" in a Project Team Scenario*

Place a "1" for all traits that represent **what you are good at** in a team or group situation.

Part "D"

_____ Making things happen
_____ Recognizing accomplishments
_____ Accepting challenges
_____ Making decisions
_____ Questioning the "status quo"
_____ Taking responsibility
_____ Causing trouble
_____ Solving problems
_____ Visualizing the future
_____ Taking charge in difficult situations

Part "I"

_____ Interacting with others
_____ Making good first impressions
_____ Being comfortable in expressing ideas
_____ Getting excited about a project
_____ Instilling enthusiasm
_____ Entertaining people
_____ Helping others
_____ Actively participating in a group
_____ Talking more than listening
_____ Not worrying about details

Part "S"

_____ Using established work procedure
_____ Staying in one physical place
_____ Showing patience
_____ Using specialized skills
_____ Concentrating on your part of project
_____ Showing loyalty
_____ Being a good listener
_____ Calming excited people
_____ Being sincere and honest
_____ Wanting to feel secure

Part "P"

_____ Following standards
_____ Concentrating on details
_____ Being supervised by others
_____ Being diplomatic
_____ Looking for accuracy
_____ Accepting decisions
_____ Criticizing performance
_____ Thinking critically
_____ Liking familiar work situations
_____ Receiving compliments for good work

SUMMARY OF YOUR DISPOSITION RESULTS: **TOTAL PART "D"** __________
TOTAL PART "I" __________
TOTAL PART "S" __________
TOTAL PART "P" __________

Chapter resource 10.3

Additional Information Relating to Personality Traits

Typical Characteristics of Personality Traits

Persons with strong **driver** traits:

May not be good at	Are motivated by	For success, need
Evaluating alternatives	Power and authority	Difficult assignments
Being careful	Prestige and challenge	To understand what others need
Working in a familiar setting	Accomplishments	To be practical
Completing research	Opportunities to advance	An occasional shock
Thinking before deciding	Direct answers	To identify with team members
Working in predictable setting	Variety of activities	To explain actions to others
Sacrificing self for others	Independence	An awareness of consequences
Patience and discipline	Diverse challenges	To relax more

Persons with strong **influencer** traits:

May not be good at	Are motivated by	For success, need
Discipline with their work	Social recognition	To use time more effectively
Gathering information	Public recognition of skills	Objectivity
Speaking directly	Freedom of expression	A democratic leader
Valuing honesty	People to talk to	Emotional control
Working independently	Nonwork group activities	A sense of urgency
Preferring things to people	Freedom from control	To be less ideological
Thinking about new ideas	Comfortable environments	To focus more on end products
Being patient	Chance to speak	To improve organizing skills

Persons with strong **specialist** traits:

May not be good at	Are motivated by	For success, need
Adapting to change	Security of situation	Conditioning prior to change
Taxing physical capacity	Stability	To use more shortcut methods
Different tasks at one time	Happy home life	To appreciate their value
Being opportunistic	Traditional procedures	Help with new tasks
Applying pressure	Sincerity by others	Supportive groups
Working in uncertain settings	Limited territory	Orders before acting
Being aggressive	Constant appreciation	Others to encourage creativity
Initiating new ideas	Identification with others	More self confidence

Persons with strong **perfectionist** traits:

May not be good at	Are motivated by	For success, need
Accepting control	Security assurances	Detailed work
Delegating work to others	Routine procedures	Careful planning
Making decisions	Safe work environment	Many explanations
Acting independently	Reassurance	Exact job descriptions
Standing their ground	Predictable circumstances	To spend less time on details
Taking unpopular stands	Being part of a group	Others to respect their value
Completing work quickly	Individual attention	Frequent feedback on progress
Initiating new ideas	Time for perfect results	To accept some imperfection

CHAPTER

11 Engineering Skills for Ethical Dilemmas

Turning Shades of Gray Situations into Black and White Solutions

Ethical Dilemma Scenario

"As engineering team leader, you are under pressure to deliver an innovative new design on time and on budget. You've developed the requirements, the design is progressing, and you are confident that the design will be a breakthrough in the market.

However, recent rumors of a competitor's new product make you doubt the superiority of the design. A colleague approaches you one day offering information that may help you gain advantage over the competitor. He has obtained a detailed presentation package that outlines the features and architecture of the competitor's emerging product. This data would allow you to revise your design to ensure it outperforms the competition.

After asking a few questions, you learn that the information was captured through questionable tactics. Furthermore, the material is marked Confidential and is covered by a Non-Disclosure Agreement executed between the competitor and one of their customers. What Will You Do?"

(James Peterman, Senior Product Manager, Strategic Projects—Tekelec)

Jim Peterman raises a very interesting question in this scenario. In our conversation with him at the end of this chapter, we will see how Jim resolved this ethical dilemma.

11.1 Introduction

Engineers impact the quality of life of others.

This case scenario introduces our topic for this chapter—ethics. Unlike many principles of engineering, ethics is difficult to structure and teach. Webster defines ethics as "pertaining to morals" and ethical as "in accordance with the rules or standards for right conduct or practice, especially the standards of a profession."

A profession is a learned occupation requiring systematic knowledge and training, and a commitment to a social good. Engineering is the creative art of applying science for the benefit of all mankind and is therefore a profession.

LEARNING OBJECTIVES

By using the information and exercises in this chapter you will be able to:

- Appreciate the need for ethical behavior.
- Recognize early signs and prepare for ethical dilemma situations.
- Use engineering tools and teams to solve ethical dilemmas.
- Evaluate potential impacts of making ethical decisions.
- Communicate recommended solutions to all stakeholders.
- Understand the impact of ethical decisions on your personal career.

Because engineering is a profession, we must consider the impact of ethics on our behavior. For example, we are ethical when we provide quality products and services. On the other hand, we are unethical if we knowingly endanger the lives of others by what we do. So, one of the most important qualities of professional engineers is to create a positive impact on the quality of life.

Ethics is especially important because what we do impacts the nontechnical world. Society has a high level of expectation for the performance and safety of our products and services. Because the general public often does not understand technology, they put their trust in engineers to design products that are safe and easy to use.

The fact that people trust us means that we have a greater responsibility to be ethical and assure personal safety and national security in all we do. Their trust also means we need to establish ethical conduct as a foundation of our career. In fact, engineering ethics is as important to good engineering practices as mathematics, physics, design skills, and other engineering fundamentals.

11.2 Preparing for Ethical Dilemmas

It is a good idea for you to start your study of ethics now. Because ethical situations and issues may appear rather quickly, the best way to be prepared to handle them is to build a foundation of ethical behavior in school, in your family, and in your community. Fortunately, you have many sources of information to help you develop this foundation.

11.2.1 Personal Code of Ethics

Your personal values are established throughout your life, starting with the impact of family and faith. Ethical values are formed when you observe actions of your parents and others. Your parents give you direction by telling and showing you what they believe is right and wrong. You learn other standards for behavior from your teachers in school and other adult models.

At first, little things may not seem to be important, but they set the stage for how you handle more important situations in your life and profession. For example, if someone gives you change for a $20 bill when you gave them $10, what do you do? The way you answer this question sets the pattern for how you may handle larger issues such as the scenario at the beginning of this chapter.

"The man of integrity walks securely, but he who takes crooked paths will be found out. Prov. 10:9

These words, written over 2000 years ago, are still valid today. Whether they are applied to our personal life or our work life, the fact is that if we compromise our integrity and sacrifice our ethics, we are no longer someone to be trusted. We may be able to deceive some, but eventually our deception will be discovered. When we cultivate a life of integrity, we'll reap the trust of our family, friends and business associates."

(Kevin F. White, PE, Manager of Engineering and Operations, Northeast Missouri Electric Power Cooperative)

As indicated in Kevin White's comments, your ability to be trusted by others depends on your personal code of ethics. This code should be based on proven standards for life and your profession. Let's review sources of information to help you further develop your code of ethics.

11.2.2 National Society of Professional Engineers' Code of Ethics

The code of ethics established by the National Society of Professional Engineers is an excellent source of information to help you define the way to practice as an engineer. This is on the Internet at: http://www.nspe.org/Ethics/CodeofEthics/index.html. The July 2007 revision of this code is shown in Chapter resource 11.1.

The preamble to this code of ethics summarizes the content of this important professional guideline as follows:

> Engineering is an important and learned profession. As members of this profession, engineers are expected to exhibit the highest standards of honesty and integrity. Engineering has a direct and vital impact on the quality of life for all people. Accordingly, the services provided by engineers require honesty, impartiality, fairness, and equity, and must be dedicated to the protection of the public health, safety, and welfare. Engineers must perform under a standard of professional behavior that requires adherence to the highest principles of ethical conduct.

"A Professional Engineer does not receive his certificate in a particular engineering discipline. However, the Canons require him or her to only practice in an area in which s/he has acquired expertise. My area of expertise is combustion. I have several times consulted with law firms on disastrous fires, which have resulted in loss of life. In each case the plaintiff's lawyer has retained a professional engineer as an expert who only had minimal training in the field.

The so-called expert created ridiculous theories as to how the fire started and why the manufacturer of the device should be held responsible. In my opinion, these experts are misusing their licenses and should not have testified outside of their knowledge. The loss of life in each case was tragic, but was not an equipment malfunction and was actually due to negligence on the part of the individuals involved. It is tempting to want to blame another party, but it should always be resisted unless there is a good scientific reason."

(Dr. Richard Cohen, PE, Associate Professor
Mechanical Engineering, Temple University)

11.2.3 Professional Societies' Codes of Ethics

Throughout this book, we emphasize the value of active membership in professional organizations. An important benefit when you are a member of these organizations is the opportunity to use their codes of ethics as an additional guide for your ethical behavior.

The following quotation by Dr. Paul Bosela highlights some of the most important fundamental principles of the American Society of Civil Engineers Code of Ethics that were adopted on September 2, 1914 and amended on July 23, 2006.

"'The safety of the public must be the first and foremost concern in the practice of civil engineering.' The American Society of Civil Engineers Code of Ethics presents the fundamental principles and canons for the ethical practice of Civil Engineering, notably stating that 'Engineers shall hold paramount the safety, health and welfare of the public and shall comply with the principles of sustainable development in the performance of their professional duties . . .' Some of the major aspects for ethical practice include 'using their knowledge and skill for the enhancement of human welfare and the environment,' 'being honest and impartial,' 'performing services only in their areas of competency,' 'uphold the honor, integrity and dignity of the engineering profession,' with 'zero-tolerance for bribery, fraud, and corruption.'"

(Dr. Paul Bosela, PE, Professor of Civil & Environmental
Engineering at Cleveland State University)

Another example of codes of conduct is shown in Chapter resource 11.2 for the American Society of Mechanical Engineers. Uniform Resource Locators (URLs) for codes of conduct of some societies are shown in Figure 11.1.

Professional organization codes of ethics are similar to standards established by the Accreditation Board for Engineering and Technology (ABET). ABET's fundamental canons, listed in Figure 11.2, are a foundation for ethical development during your academic experiences.

American Institute of Chemical Engineers
http://www.ehow.com/facts_6823508_code-american-institute-chemical-engineers.html

American Society of Civil Engineers
http://www.asce.org/Leadership-and-Management/Ethics/Code-of-Ethics/

American Society of Mechanical Engineers
http://files.asme.org/asmeorg/governance/3675.pdf

Association for Computing Machinery
http://courses.cs.vt.edu/cs3604/lib/WorldCodes/ACM.Code.1992.html

IEEE
http://www.ieee.org/about/corporate/governance/p7-8.html

Figure 11.1
Sample professional society codes of ethics.

ABET FUNDAMENTAL CANONS

1. Engineers shall hold paramount the safety, health, and welfare of the public in the performance of their professional duties
2. Engineers shall perform services only in the areas of their competence
3. Engineers shall issue public statements only in an objective and truthful manner
4. Engineers shall act in professional matters for each employer or client as faithful agents or trustees, and shall avoid conflicts of interest
5. Engineers shall build their professional reputation on the merit of their services and shall not compete unfairly with others
6. Engineers shall act in such a manner as to uphold and enhance the honor, integrity, and dignity of the profession
7. Engineers shall continue their professional development throughout their careers and shall provide opportunities for the professional development of those engineers under their supervision

Figure 11.2
ABET fundamental canons.

11.2.4 Corporate Codes of Ethics

Most corporations include ethical procedures in their formal or informal operating practices. These may be found in corporate objectives and policies, or they may be more informal in the established culture of the company. Figure 11.3 identifies typical topics in many corporate codes of ethics.

It is a good idea for you to discuss ethical attitudes of potential companies during job interviews. This can be accomplished by asking for a copy of their code of ethics or by asking questions that refer to their attitude about the impact of their product or service on customers.

To achieve ethical goals, many corporations establish procedures for employees to discuss ethical situations with higher management. This typically includes talking to your supervisor, or in some cases, higher levels of management. Some procedures may also include talking to an individual in the Human Resources Department or an outside consultant.

Following the established corporate code of ethics is important in all job situations. These help you define guidelines for making decisions about the quality of your work and your relationships to other employees and customers.

11.3 Recognizing Early Signs of Ethical Dilemmas

The third paragraph of the ethical scenario outlined at the start of this chapter indicates an ethical situation is likely involved. Some of these statements are red flags of unusual

TYPICAL TOPICS
IN CORPORATE CODES OF ETHICS

- Product quality
- Product support
- Customer relations
- Accountability with company funds
- Gifts from suppliers
- Trade secrets
- Environmental impacts
- Public health and safety
- International operations
- Travel and company representation
- Information relating to competitors

Figure 11.3
Corporate codes of ethics topics.

ENGINEERING PROBLEM-SOLVING PROCESS

1. Define the problem
2. Determine objectives
3. Identify alternative solutions
4. Identify constraints
5. Select solution
6. Test solution for potential impacts
7. Document and communicate solution

Figure 11.4
Problem-solving process.

circumstances or activities that cause you to stop and consider if the suggestions of the colleague are ethical and should be used. There are often early signs when things are unethical and engineers have the responsibility to evaluate these signs.

Look for early warning signs of unethical behavior.

When early signs of a potential ethical situation are noted, the next step is to record information. This builds a foundation on which you can decide if there is a problem, and if so, what you should do next. And this brings us to the most important part of handling ethical dilemmas—using engineering tools to solve them.

11.4 Using Engineering Tools for Ethical Dilemmas

As you investigate and document a potential ethical situation, you will discover that many of the issues are shades of gray with different impacts on different people. This makes your ability to find the right solution a little more challenging. Your goal in addressing an ethical dilemma should be to change these shades of gray situations to black and white solutions. So, how can you do that?

In many of your classes, you are learning how to solve engineering problems by following a step-by-step process shown in Figure 11.4. This same approach can be used to solve ethical problems. You can face an ethical situation with confidence because, with your engineering skills, you will know where to begin, how to proceed, and how to evaluate results of your solution.

11.4.1 Define the Problem

In solving a technical engineering problem, the most important step is for you to first identify all known quantities. This approach is also the most important step in addressing an ethical situation.

When something appears to be unethical based on comments from coworkers or casual observation, this is the time to look for facts. As you identify details of the situation, avoid stating or implying that this is unethical. It is risky to make hasty judgments about an ethical situation without first determining all the facts.

A decision about ethics should only be made after all information has been gathered. If you make false accusations, your reputation and career may be damaged.

Start your investigation of a potential unethical situation by obtaining quantitative information. You can do this by requesting information from others involved to ensure the situation is being properly assessed. Additional guidance can be solicited from friends and coworkers who can be trusted to keep the evaluation process confidential. Be sure to document detailed information.

11.4.2 Determine Objectives

When it appears something is unethical, based on the facts of early investigations, the next step is to determine if there is a reason for you to do something about it. The answer to that question is determined by the extent of the unethical act.

For example, is it important for you to disclose a student who is cheating in class? The answer may be yes if this is on a test and could impact your grade and the grades of other students. However, you might first discuss this privately with the student and suggest this is not in his or her best long-term interest.

If one of your laboratory team members is unethical and uses information from stolen laboratory reports, this might also reflect on you and other team members. In this case, you really should discuss this with your teammate and then with the instructor, if necessary.

> The decision to be involved in an ethical dilemma is based on codes of conduct.

As a student, your decision about becoming involved in ethical dilemmas can be determined by your personal code of ethics. You will be more successful in handling early ethical dilemmas if your personal code is based on professional codes. As a practicing engineer, decisions for involvement should always be based on professional and corporate codes of ethics.

An analysis, similar to the ones we discussed for student unethical situations, can be used when you consider getting involved in an ethical situation in your working career. Start by asking yourself how important is the ethical dilemma.

For example, are people taking pencils home for their children to use in school or are they giving trade secrets to a competitor? Does the unethical practice impact you, your company, or customers? Your decision about getting involved in ethical situations is usually based on potential impacts on you, your family, and your company.

The first step in most company ethical policies is to discuss situations with your supervisor. However, if your supervisor is involved in the unethical situation, you have a major challenge. The objective of your involvement in this case usually will be determined by weighing potential risks of talking to a third party or bypassing your supervisor and talking to someone in the next level of management.

11.4.3 Identify Alternative Solutions

When you decide to address an ethical dilemma, evaluate options for your approach to find a solution. If only one person is unethical, you might start by encouraging him or her to change. If many are involved, your objective may be to use the corporate process established to handle ethical problems.

For more complex ethical dilemmas, it is appropriate to form a team of associates to review as many options as possible and consider the consequences of each. Typically, this will reveal negative and positive impacts on various individuals or groups, and you can find the best solution for most stakeholders by a careful review of these options.

By using the process introduced in the chapter on team building, potential alternative solutions will be expanded because more team members will be contributing ideas. Group synergy can result in finding more options and selecting better final recommendations.

Another reason for using a team approach is that there is less personal risk when solving ethical dilemmas. The recommendations of several people are usually considered more viable than those offered by just one person. If recommendations are unpopular, the negative impact on individual team members is reduced when presented as a team solution.

11.4.4 Identify Constraints

Because alternative solutions include various constraints, it may be difficult to select the "best" solution. Ethical dilemmas are usually not black or white but different shades of gray. You can define some constraints by reviewing records made during the discovery process. Remember, your major objective is to change something that appears to be shades of gray to black and white options and then select a solution based on professional codes of ethics.

You can identify other constraints as you investigate the details of alternative solutions. As a student, if you suggest a laboratory team member should stop plagiarizing reports, you may not have the support of other team members. As a practicing engineer, you may be working for a company that has limited funds to cope with environmental impacts of a project. Most options have related constraints, so identify and include constraints in the evaluation process when seeking solutions.

11.4.5 Select Solution

And that brings us to the most difficult part of resolving ethical dilemmas—finding the right solution. You can select appropriate solutions by using documented facts, evaluating options and constraints, and holding discussions with advisors and team members.

Impacts on various stakeholders should also be part of the equation. Your ability to have others accept and support your decisions is often based on how it impacts them personally. In all cases, solutions should be based on personal, professional, and company ethical codes.

11.4.6 Test Solution for Potential Impacts

Before making suggested solutions public, it is a good idea to test potential impacts on all involved. This can be done within a team, by talking to trusted friends, and by discussing with supervision. Decisions have different impacts on different stakeholders. If one person is unethical, the implementation of a solution to correct the problem may hurt his or her career, but might save the careers of many others who work for the company.

> Test solutions for possible impacts before making final decisions.

Stakeholders include customers as well as the company. If a product is unsafe, it could injure those who use it. In addition to causing major injury to customers, this could also result in costly product recalls or lawsuits. This in turn could impact all employees.

Other important stakeholders are your family, as the discussion in Figure 11.5 shows. Although they are not directly involved in your work, they are impacted by what you do. To include their input, discuss potential impacts relating to ethical solutions with your family before finalizing your decisions.

11.4.7 Document and Communicate Solution

It is important to document information from start to finish when addressing an ethical dilemma. This provides facts on which you can make better decisions, and is an important resource when you communicate recommendations.

The final step in solving an engineering problem is to communicate the answer to your instructor or supervisor. The same holds true for solutions to ethical problems. Your effectiveness to communicate the solution, and the reasons for it, will greatly

Figure 11.5
Family discussions of ethical dilemma options.

dictate how others will accept and support it. So, time used in documenting and communicating your recommendations is a good investment and may be the difference in your solution being accepted or rejected.

11.5 Impact of Ethics on Career Success

As indicated in the introduction to this chapter, the nontechnical world relies on engineers to provide quality products and services. When this trust is violated by unethical conduct, there are consequences. Your company and your personal career can experience very negative impacts when ethics are not an integral part of everything you do.

11.5.1 Long-Term Consequences of Ethics

The consequences of some unethical activities may not at first appear to be harmful. Examples of taking a pencil home or using the company copy machine for personal reasons appear to be insignificant. After all, some people may think the company is big, has a lot of money, and no one will notice these actions.

But the difficulty with petty unethical actions is that they can develop into something much more significant. Then the consequences can be damaging to the company, to the person, and to his or her career. The idea of stealing a pencil is really just as unethical as stealing trade secrets to share with a competitor. In both cases, the property belongs to someone else.

Just as your career started the day you walked in the door as a freshman, your ethical attitude should also start early. A good way to establish ethical behavior is to incorporate the ABET Fundamental Canons and professional codes of ethics into your personal code of ethics.

Continue your focus on ethics by basing your behavior on professional codes as you start your working career. Because most companies desire a positive ethical image for long-term stability, you need to include ethical behavior when working for a company. Ethical behavior is one of the best methods to deliver quality work and products and to increase the potential level of your personal success.

Unethical behavior in an organization eventually will be discovered. If you are the one being unethical, this can interfere with your future opportunities within the company or result in termination, as indicated in Figure 11.6. Though it might not be apparent in the beginning, unethical behavior is risky and can result in negative impacts on your career.

11.5.2 Value of a Position of Strength

When you are involved in an ethical dilemma, you may experience personal risks, and you need to be prepared for possible results. Your recommendations may be perceived as "blowing the whistle" on an unethical situation, and could result in you being terminated by the company.

So, how can you prepare for possible negative impacts when facing an ethical dilemma? The best answer to this question is for you to operate from a position of strength. This starts with a history of your loyalty to the company and your quality

Figure 11.6
Potential results of unethical behavior.

work. Your position becomes stronger when your supervisor and others are aware of your contributions to the company's success.

The second way to be in a strong position is to have a well-designed career plan that helps you anticipate potential detours and be ready to take advantage of unexpected opportunities. We will discuss career plans in detail in the chapter on career management.

To cope with potential negative impacts relating to solving an ethical problem, be proactive and not reactive. Then, you will be much more in control of your future, and can use detours to your advantage.

A CONVERSATION WITH JIM PETERMAN

Jim, how did you handle the ethical dilemma introduced in the beginning of this chapter?

Answer:

"Basic decision making and objective analysis is the key to addressing this dilemma. What are the pros and cons of using the data? While the benefits initially seem worth any price, the realization that the data is effectively stolen brings many more disadvantages. There could be legal repercussions to using the data including issues related to the NDA as well as potential violation of Intellectual Property Rights. Our personal integrity and that of our company were also at great risk. Communication and interpersonal skills were also required to discuss our decision with the colleague who offered the information and to explain our rationale for not accepting the data.

In summary, the project remained on-track without any design iterations and it was released on schedule. While competitor's products did emerge that were similar, our ability to meet schedules and demonstrate integrity helped mature and improve relationships with key customers that eventually led to long-term purchasing agreements."

What nontechnical skills did you use in this situation?

Answer:

"It is important to establish personal discipline to prevent reacting with rashness to unexpected situations. All decisions, especially nontechnical ones, require careful assessment, so you must exercise the discipline to make time to consider the situation. Knowing that difficult situations will arise, you should identify experienced individuals who demonstrate wisdom who can serve as trusted advisors and mentors. Select persons who reflect integrity in all situations."

What advice would you share with students?

Answer:

"Consider nontechnical skills that are required for your job and give equal energy to developing them. Understanding ethical issues prior to facing difficult dilemmas will help your preparation. Review IEEE's Code of Ethics and your company's Rules of Business Conduct. Also identify your company's ethics contacts, who usually offer anonymity for discussing or reporting ethical issues. Identify one or more trusted resources that you can use for advice and counsel. Finally, realize that your reputation for engineering excellence and ethical integrity is your greatest personal asset. Tarnishing it through a rush to judgment when an ethical dilemma arises may have lasting negative effects on your career."

(James Peterman, Senior Product Manager, Strategic Projects—Tekelec)

11.6 Conclusion

As we have discussed in this chapter, black and white solutions need to be developed for ethical situations that first appear to be shades of gray. In such dilemmas, engineers play a major role because they can use proven codes of ethics to find the best answer. In all cases, we need to ask if it should be done in addition to can it be done.

In making ethical decisions, we need to be sensitive to how our decisions will impact others. Many options will have different, but equally important, results. Various interest groups may be impacted in important ways.

Although ethical behavior is based on personal values, you need to use professional guidelines to help you make ethical decisions. These decisions may not be easy and the results can have negative impacts on your company, your career, and your family. Because others will be affected, you should discuss your options with them before making your final decision on what you will do when faced with an ethical decision.

In the competitive business world, there is often the temptation to focus on short-term profits. While profits are important for companies to stay in business, to be professional, you should accept the responsibility of ethical behavior and the relationship of your work to the overall society in which you live. The time for you to develop ethical attitudes and skills to handle future ethical situations is now.

END OF CHAPTER REVIEW QUIZ

Select the most appropriate answer to the following statements.

1. The best way to develop ethical behavior is to
   - **a.** Ask your trusted friends how to handle an unethical situation.
   - **b.** Read guidelines published by professional societies.
   - **c.** Follow personal, professional, and company ethical guidelines.
   - **d.** Be ethical in your work assignments.
2. You can usually see early signs of unethical behavior by
   - **a.** Gathering facts and comparing them to ethical standards.
   - **b.** Just using common sense.
   - **c.** Asking people why they are doing something suspicious.
   - **d.** Guilty looks on someone's face.
3. The most important guideline for being ethical is
   - **a.** What your parents taught you about what is right and wrong.
   - **b.** What the company you work for establishes as ethical codes.
   - **c.** What your supervisor and peers tell you what is right.
   - **d.** What professional organizations publish about ethics.
4. When a possible unethical dilemma is identified the first thing to do is
   - **a.** Tell you instructor or supervisor there is a problem.
   - **b.** Form a team and start looking for a solution.
   - **c.** Gather and document as many facts as possible.
   - **d.** Go to the local newspaper and ask them to investigate the situation.
5. The most important value of using a team to find a solution to an ethical problem is
   - **a.** So you can blame other team members if people don't like the solution.
   - **b.** To find more constraints to possible solutions.
   - **c.** To delay making a decision and hope the problem will go away.
   - **d.** To develop the best solution by using the synergy of group discussions.
6. Before making a final recommendation to an ethical dilemma, the best thing is to
   - **a.** Document everything about the situation.
   - **b.** Test possible impacts on all stakeholders.
   - **c.** Determine the objective of your involvement in trying to find a solution.
   - **d.** Ask your instructor or supervisor if they like your recommendation.
7. The greatest risk in getting involved in solving an ethical dilemma is
   - **a.** You may lose your job and have a reputation for "whistle blowing."
   - **b.** The unethical person may find a way to get even.
   - **c.** You might waste a lot of your personal time.
   - **d.** Your peers may never speak to you again.

8. The biggest problem in being unethical in small things is
- **a.** You might not get a promotion if someone knows what you did.
- **b.** You could develop a pattern that leads to much more serious unethical actions.
- **c.** You are taking a chance that someone will find out.
- **d.** None of the above.

9. A major potential long-term impact of being unethical is
- **a.** You will feel guilty and lose self-esteem.
- **b.** You might damage the reputation of your employer.
- **c.** You can be terminated by your employer.
- **d.** All of the above.

10. The best way to be prepared for potential impacts when solving an ethical dilemma is
- **a.** To have a career plan that prepares you for changing jobs if necessary.
- **b.** To have a strong friendship with your supervisor.
- **c.** To have family support in case you are terminated.
- **d.** Not having any debt so you don't lose your home in case you are terminated.

EXERCISES TO DEVELOP AND ENHANCE YOUR SKILL SET

Exercise 11.1 Ethical dilemma exercise

Select one of the three ethical dilemmas listed in Chapter resource 11.3, Ethical dilemma examples, and use the **ETHICAL DILEMMA REPORT** to answer questions and provide written comments and your recommended solution to the dilemma.

Include a paragraph under **ETHICAL DILEMMA EXERCISE ASSESSMENT** to report your evaluation of this exercise relating to this ethical dilemma.

Take your report to the university writing center and request their staff to review and comment on your writing skills. Then, add a final paragraph under **UNIVERSITY WRITING CENTER CONTACT AND EXPERIENCE ASSESSMENT** to report the name of the person who reviewed your report and discuss the experience of working with the center for this assignment.

The due date for your report will be established by your instructor. Hand in your written report at the start of class and send an electronic copy to your instructor on the due date.

Ethical Dilemma Report

By ______________________________ (Your Name)

Ethical Dilemma Evaluation

1. Ethical dilemma summary

Summarize the major points of the dilemma in one paragraph.

2. Solution options

Develop various options for a solution to this dilemma and discuss them using one paragraph for each option.

3. Potential impact of solution options

Define potential positive and negative impacts for each solution option defined in the second part of this report and discuss them using one paragraph for each option.

4. Recommended solution

Select the solution you recommend and use one paragraph to discuss the reasons for recommending this solution.

Ethical Dilemma Exercise Assessment

Use one paragraph to discuss your experience in completing this assignment and your impression of the value of participating in this exercise.

University Writing Center Contact and Experience Assessment

Use one or more paragraphs to report the name of the person you contacted in the University Writing Center, your experience in working with the center and how this experience will help you in future writing assignments.

Exercise 11.2 Ethical dilemma team exercise

Review the ethical dilemma in Chapter resource 11.4, SAMPLE NSPE ETHICAL CASE STUDIES, assigned by your instructor. Conduct a team discussion as assigned and prepare your individual **ETHICAL DILEMMA REPORT** to answer questions, and provide written comments relating to the team discussion and team recommended solution to the dilemma.

Include a paragraph under **ETHICAL DILEMMA EXERCISE ASSESSMENT** to report your evaluation of this exercise relating to this ethical dilemma.

Take your report to the university writing center and request their staff to review and comment on your writing skills. Then, add a final paragraph under **UNIVERSITY WRITING CENTER CONTACT AND EXPERIENCE ASSESSMENT** to report the name of the person who reviewed your report and discuss the experience of working with the center for this assignment.

The due date for your report will be established by your instructor. Hand in your written report at the start of class and send an electronic copy to your instructor on the due date.

Ethical Dilemma Report

By ______(Your Name)______

Ethical Dilemma Evaluation

1. Ethical dilemma summary

Summarize the major points of the dilemma in one paragraph.

2. Solution options

Develop various options for a solution to this dilemma and discuss them using one paragraph for each option.

3. Potential impact of solution options

Define potential positive and negative impacts for each solution option defined in the second part of this report and discuss them using one paragraph for each option.

4. Recommended solution

Select the solution you recommend and use one paragraph to discuss the reasons for recommending this solution.

Ethical Dilemma Exercise Assessment

Use one paragraph to discuss your experience in completing this assignment and your impression of the value of participating in this exercise.

University Writing Center Contact and Experience Assessment

Use one or more paragraphs to report the name of the person you contacted in the University Writing Center, your experience in working with the center, and how this experience will help you in future writing assignments.

Exercise 11.3 Study group ethical dilemma discussion

If your instructor does not assign Exercise 11.2, you can enhance your team building and ethical skills by selecting one of the NSPE Ethical Cases from the Chapter resource 11.4 and discussing this case with a study group.

END OF CHAPTER RESOURCES

Chapter resource 11.1 National Society of Professional Engineers Code of Ethics for Engineers

National Society of Professional Engineers Code of Ethics for Engineers

As Revised July 2007

Preamble

Engineering is an important and learned profession. As members of this profession, engineers are expected to exhibit the highest standards of honesty and integrity. Engineering has a direct and vital impact on the quality of life for all people. Accordingly, the services provided by engineers require honesty, impartiality, fairness, and equity, and must be dedicated to the protection of the public health, safety, and welfare. Engineers must perform under a standard of professional behavior that requires adherence to the highest principles of ethical conduct.

I. Fundamental Canons

Engineers, in the fulfillment of their professional duties, shall:

1. Hold paramount the safety, health, and welfare of the public.
2. Perform services only in areas of their competence.
3. Issue public statements only in an objective and truthful manner.
4. Act for each employer or client as faithful agents or trustees.
5. Avoid deceptive acts.
6. Conduct themselves honorably, responsibly, ethically, and lawfully so as to enhance the honor, reputation, and usefulness of the profession.

II. Rules of Practice

1. Engineers shall hold paramount the safety, health, and welfare of the public.
   a. If engineers' judgment is overruled under circumstances that endanger life or property, they shall notify their employer or client and such other authority as may be appropriate.
   b. Engineers shall approve only those engineering documents that are in conformity with applicable standards.
   c. Engineers shall not reveal facts, data, or information without the prior consent of the client or employer except as authorized or required by law or this Code.
   d. Engineers shall not permit the use of their name or associate in business ventures with any person or firm that they believe is engaged in fraudulent or dishonest enterprise.
   e. Engineers shall not aid or abet the unlawful practice of engineering by a person or firm.
   f. Engineers having knowledge of any alleged violation of this Code shall report thereon to appropriate professional bodies and, when relevant, also to public authorities, and cooperate with the proper authorities in furnishing such information or assistance as may be required.

2. Engineers shall perform services only in the areas of their competence.

a. Engineers shall undertake assignments only when qualified by education or experience in the specific technical fields involved.

b. Engineers shall not affix their signatures to any plans or documents dealing with subject matter in which they lack competence, nor to any plan or document not prepared under their direction and control.

c. Engineers may accept assignments and assume responsibility for coordination of an entire project and sign and seal the engineering documents for the entire project, provided that each technical segment is signed and sealed only by the qualified engineers who prepared the segment.

3. Engineers shall issue public statements only in an objective and truthful manner.

a. Engineers shall be objective and truthful in professional reports, statements, or testimony. They shall include all relevant and pertinent information in such reports, statements, or testimony, which should bear the date indicating when it was current.

b. Engineers may express publicly technical opinions that are founded upon knowledge of the facts and competence in the subject matter.

c. Engineers shall issue no statements, criticisms, or arguments on technical matters that are inspired or paid for by interested parties, unless they have prefaced their comments by explicitly identifying the interested parties on whose behalf they are speaking, and by revealing the existence of any interest the engineers may have in the matters.

4. Engineers shall act for each employer or client as faithful agents or trustees.

a. Engineers shall disclose all known or potential conflicts of interest that could influence or appear to influence their judgment or the quality of their services.

b. Engineers shall not accept compensation, financial or otherwise, from more than one party for services on the same project, or for services pertaining to the same project, unless the circumstances are fully disclosed and agreed to by all interested parties.

c. Engineers shall not solicit or accept financial or other valuable consideration, directly or indirectly, from outside agents in connection with the work for which they are responsible.

d. Engineers in public service as members, advisors, or employees of a governmental or quasi-governmental body or department shall not participate in decisions with respect to services solicited or provided by them or their organizations in private or public engineering practice.

e. Engineers shall not solicit or accept a contract from a governmental body on which a principal or officer of their organization serves as a member.

5. Engineers shall avoid deceptive acts.

a. Engineers shall not falsify their qualifications or permit misrepresentation of their or their associates' qualifications. They shall not misrepresent or exaggerate their responsibility in or for the subject matter of prior assignments. Brochures or other presentations incident to the solicitation of employment shall not misrepresent pertinent facts concerning employers, employees, associates, joint ventures, or past accomplishments.

b. Engineers shall not offer, give, solicit, or receive, either directly or indirectly, any contribution to influence the award of a contract by public authority, or which may be reasonably construed by the public as having the effect or intent of influencing the awarding of a contract. They shall not offer any gift or other valuable consideration in order to secure work. They shall not pay a commission, percentage, or brokerage fee in order to secure work, except to a bona fide employee or bona fide established commercial or marketing agencies retained by them.

III. Professional Obligations

1. Engineers shall be guided in all their relations by the highest standards of honesty and integrity.
   a. Engineers shall acknowledge their errors and shall not distort or alter the facts.
   b. Engineers shall advise their clients or employers when they believe a project will not be successful.
   c. Engineers shall not accept outside employment to the detriment of their regular work or interest. Before accepting any outside engineering employment, they will notify their employers.
   d. Engineers shall not attempt to attract an engineer from another employer by false or misleading pretenses.
   e. Engineers shall not promote their own interest at the expense of the dignity and integrity of the profession.
2. Engineers shall at all times strive to serve the public interest.
   a. Engineers are encouraged to participate in civic affairs; career guidance for youths; and work for the advancement of the safety, health, and well-being of their community.
   b. Engineers shall not complete, sign, or seal plans and/or specifications that are not in conformity with applicable engineering standards. If the client or employer insists on such unprofessional conduct, they shall notify the proper authorities and withdraw from further service on the project.
   c. Engineers are encouraged to extend public knowledge and appreciation of engineering and its achievements.
   d. Engineers are encouraged to adhere to the principles of sustainable development in order to protect the environment for future generations.
3. Engineers shall avoid all conduct or practice that deceives the public.
   a. Engineers shall avoid the use of statements containing a material misrepresentation of fact or omitting a material fact.
   b. Consistent with the foregoing, engineers may advertise for recruitment of personnel.
   c. Consistent with the foregoing, engineers may prepare articles for the lay or technical press, but such articles shall not imply credit to the author for work performed by others.
4. Engineers shall not disclose, without consent, confidential information concerning the business affairs or technical processes of any present or former client or employer, or public body on which they serve.
   a. Engineers shall not, without the consent of all interested parties, promote or arrange for new employment or practice in connection with a specific project for which the engineer has gained particular and specialized knowledge.
   b. Engineers shall not, without the consent of all interested parties, participate in or represent an adversary interest in connection with a specific project or proceeding in which the engineer has gained particular specialized knowledge on behalf of a former client or employer.
5. Engineers shall not be influenced in their professional duties by conflicting interests.
   a. Engineers shall not accept financial or other considerations, including free engineering designs, from material or equipment suppliers for specifying their product.
   b. Engineers shall not accept commissions or allowances, directly or indirectly, from contractors or other parties dealing with clients or employers of the engineer in connection with work for which the engineer is responsible.

6. Engineers shall not attempt to obtain employment or advancement or professional engagements by untruthfully criticizing other engineers, or by other improper or questionable methods.
   - **a.** Engineers shall not request, propose, or accept a commission on a contingent basis under circumstances in which their judgment may be compromised.
   - **b.** Engineers in salaried positions shall accept part-time engineering work only to the extent consistent with policies of the employer and in accordance with ethical considerations.
   - **c.** Engineers shall not, without consent, use equipment, supplies, laboratory, or office facilities of an employer to carry on outside private practice.
7. Engineers shall not attempt to injure, maliciously or falsely, directly or indirectly, the professional reputation, prospects, practice, or employment of other engineers. Engineers who believe others are guilty of unethical or illegal practice shall present such information to the proper authority for action.
   - **a.** Engineers in private practice shall not review the work of another engineer for the same client, except with the knowledge of such engineer, or unless the connection of such engineer with the work has been terminated.
   - **b.** Engineers in governmental, industrial, or educational employ are entitled to review and evaluate the work of other engineers when so required by their employment duties.
   - **c.** Engineers in sales or industrial employ are entitled to make engineering comparisons of represented products with products of other suppliers.
8. Engineers shall accept personal responsibility for their professional activities, provided, however, that engineers may seek indemnification for services arising out of their practice for other than gross negligence, where the engineer's interests cannot otherwise be protected.
   - **a.** Engineers shall conform with state registration laws in the practice of engineering.
   - **b.** Engineers shall not use association with a nonengineer, a corporation, or partnership as a "cloak" for unethical acts.
9. Engineers shall give credit for engineering work to those to whom credit is due, and will recognize the proprietary interests of others.
   - **a.** Engineers shall, whenever possible, name the person or persons who may be individually responsible for designs, inventions, writings, or other accomplishments.
   - **b.** Engineers using designs supplied by a client recognize that the designs remain the property of the client and may not be duplicated by the engineer for others without express permission.
   - **c.** Engineers, before undertaking work for others in connection with which the engineer may make improvements, plans, designs, inventions, or other records that may justify copyrights or patents, should enter into a positive agreement regarding ownership.
   - **d.** Engineers' designs, data, records, and notes referring exclusively to an employer's work are the employer's property. The employer should indemnify the engineer for use of the information for any purpose other than the original purpose.
   - **e.** Engineers shall continue their professional development throughout their careers and should keep current in their specialty fields by engaging in professional practice, participating in continuing education courses, reading in the technical literature, and attending professional meetings and seminars.

Chapter resource 11.2 American Society of Mechanical Engineers Code of Ethics of Engineers

ASME Code of Ethics of Engineers

As Revised April 2009

ASME requires ethical practice by each of its members and has adopted the following Code of Ethics of Engineers as referenced in the ASME Constitution, Article C2.1.1.

The Fundamental Principles

Engineers uphold and advance the integrity, honor and dignity of the engineering profession by:

I. using their knowledge and skill for the enhancement of human welfare;

II. being honest and impartial, and serving with fidelity their clients (including their employers) and the public; and

III. striving to increase the competence and prestige of the engineering profession.

The Fundamental Canons

1. Engineers shall hold paramount the safety, health and welfare of the public in the performance of their professional duties.
2. Engineers shall perform services only in the areas of their competence; they shall build their professional reputation on the merit of their services and shall not compete unfairly with others.
3. Engineers shall continue their professional development throughout their careers and shall provide opportunities for the professional and ethical development of those engineers under their supervision.
4. Engineers shall act in professional matters for each employer or client as faithful agents or trustees, and shall avoid conflicts of interest or the appearance of conflicts of interest.
5. Engineers shall respect the proprietary information and intellectual property rights of others, including charitable organizations and professional societies in the engineering field.
6. Engineers shall associate only with reputable persons or organizations.
7. Engineers shall issue public statements only in an objective and truthful manner and shall avoid any conduct which brings discredit upon the profession.
8. Engineers shall consider environmental impact and sustainable development in the performance of their professional duties.
9. Engineers shall not seek ethical sanction against another engineer unless there is good reason to do so under the relevant codes, policies and procedures governing that engineer's ethical conduct.
10. Engineers who are members of the Society shall endeavor to abide by the Constitution, By-Laws and Policies of the Society, and they shall disclose knowledge of any matter involving another member's alleged violation of this Code of Ethics or the Society's Conflicts of Interest Policy in a prompt, complete and truthful manner to the chair of the Committee on Ethical Standards and Review.

Chapter resource 11.3 Ethical dilemma examples

Deciding What Belongs to a Company and What Belongs to You

During lunch breaks from work, you and one of your coworkers developed a new spreadsheet program on the computer in the office. It is powerful, yet easier to use than similar programs available to purchase. Because this program is so unique, your friend wants to market it.

This is a very attractive thought because you have many college loans to repay and you are confident this could result in a large profit for both of you. However, you developed this program using company equipment during time in the company office. What should you do?

Moving Software Design from One Company to Another

Upon graduation as a computer engineer, you worked for a small computer firm that specializes in developing software for management tasks. You were the primary contributor in designing an innovative software system for customer services.

Now, you are working for a much larger computer firm. You realize that by making a few minor changes in the innovative software system you designed at the small computer firm you could use this to simplify and significantly enhance your new tasks at the larger firm. Because you no longer work for the firm where you developed this software, should you use a slightly different version of this in your present position to help your employer by being more productive?

Dialysis Equipment for Different Economic Situations

As a design engineer for a company that supplies dialysis equipment for worldwide use, you know resources are limited for third-world countries compared to economically developed countries. Secondary infection is a risk in dialysis and patients with kidney disease could experience long-term injury and increased risk of death if infected. You have a choice between three designs:

- The first design will have a secondary infection rate of 1 in 1,000.
- The second design will have a secondary infection rate of 1 in 1,000,000 and cost 10 times as much as the first design.
- The third design will have a secondary infection rate of 1 in 5,000,000 but will cost 100 times the amount of the first design.

Questions? Which design will you choose? Would it make a difference if the design was primarily for the economically developed countries or for the third-world countries?

Chapter resource 11.4 Sample NSPE ethical case studies

The sources of information in this Chapter Resource are ethics cases listed on the website for the National Society of Professional Engineers http://www.niee.org/cases/.

This information is for educational purposes only. It may be reprinted without further permission provided that this statement is included before or after the text of the case AND that appropriate attribution is provided to the National Society of Professional Engineers' Board of Ethical Review.

CASE STUDY 1

Employment Information Gained from ABET Visitation

Case Scenario:

Engineer A is an engineering educator who serves as an evaluator on an Accreditation Board for Engineering and Technology (ABET) visitation team reviewing an engineering program at State X University. Because Engineer B is considering leaving State X University, State X University may have an opening in the position of chair of its chemical engineering department. State X University has not advertised or announced the opening.

Engineer A visits State X University with an ABET visitation team in October. During the ABET visit and interviews with the dean, department chairs, and engineering faculty, it becomes apparent to Engineer A that Engineer B might depart and a chemical engineering chair position at State X University might open.

Engineer A completes her evaluation and has no further influence on the report or the final action by ABET. In June, she formally learns of the chair position opening, applies for the position, is selected by a search committee, and accepts the appointment as chair of the State X University chemical engineering department.

Question:

Was it ethical for Engineer A to apply for the position of chair at State X University?

CASE STUDY 2

Use of Alleged Hazardous Material in a Processing Facility

Case Scenario:

Engineer A is a graduate engineer in a manufacturing facility that uses toxic chemicals in its processing operations. His job has nothing to do with the use and control of these materials.

A chemical called "MegaX" is used at the site. Recent stories in the news have reported alleged immediate and long-term human genetic hazards from inhalation of or other contact with MegaX. The news items are based on findings from laboratory experiments, which were done on mice, by a graduate student at a well-respected university's physiology department. Other scientists have neither confirmed nor refuted the experimental findings. Federal and local governments have not made official pronouncements on the subject.

Several colleagues outside of the company have approached Engineer A on the subject and ask Engineer A to "do something" to eliminate the use of MegaX at the processing facility.

Engineer A mentions this concern to her manager who tells Engineer A, "Don't worry, we have an Industrial Safety Specialist who handles that."

Two months elapse and MegaX is still used in the factory. The controversy in the press continues, but since there is no further scientific evidence pro or con in the matter, the issues remain unresolved. The use of the chemical in the processing facility has increased and now more workers are exposed daily to the substance than was the case two months ago.

Question:

Does Engineer A have an obligation to take further action under the facts and circumstances?

CASE STUDY 3

Offer of Employment by Vendor

Case Scenario:

Upon graduation from an ABET/EAC-accredited civil engineering program, Engineer A is employed by U&I Construction Co., which is owned and operated by Engineer B and Engineer C, both licensed professional engineers. Engineer A is soon delegated the responsibility of preparing bills of materials for construction designs that include appropriate allowance for waste and negotiations of material procurement with suppliers.

Engineer A negotiates quantity, schedule, specifications, and price, and then submits a recommendation to his highly experienced, nondegreed supervisor to arrange for appropriate company approval authority for the procurement contract if the financial commitment to a supplier on a project exceeds $250. After two years, Engineer A expresses concern to his supervisor that his job seems repetitive and lacks the variety of experiences and challenges that draw on the breadth of his education.

Engineer A is informed that he is providing an essential service to the company with exceptional proficiency, for which he seems very well paid, and that he will be considered for opportunities should they become available—if a replacement to cover his current activities can be found. Engineer A's financial authority is increased to $500 for any one supplier of the project.

Another year passes and Engineer A is still performing the same level of assignments. He has developed a highly respected reputation for knowledge, fairness, and integrity among the suppliers of U&I Construction Co. Engineer D, an employee of ACE Supplies, a frequent supplier to U&I, has developed a working relationship with Engineer A. When ACE has an opening for a civil engineer, Engineer D tells Engineer A about it. Engineer A interviews for the position and, after an evaluation period, Engineer A receives an offer of employment with ACE.

The offer letter states that ACE is "looking forward to having Engineer A on its team commencing on a mutually agreed upon date . . . that Engineer A is not an employee of ACE until Engineer A physically reports to work at ACE's facilities, executes patent and proprietary information agreements at that time, and that the employer's physician confirms that Engineer A has no preexisting health condition that would prevent Engineer A from performing the requirements of the position."

In a subsequent discussion with Engineer A, Engineer D mentions that the position was one that the ACE Vice President of Engineering has the prerogative of filling, but on occasion the ACE CEO has eliminated the position opening even with outstanding offers pending until business conditions improved or when a major customer had expressed displeasure with the hiring of one of its employees.

Engineer A submits his resignation with a customary two-week notice to U&I. Engineer A's supervisor, Engineer E, is disturbed by the resignation and expresses a desire that Engineer A stay with U&I, saying that if Engineer E could prevent his leaving he would. Engineer A insists that his decision is firm. Engineer A is not asked and does not believe it is in his interests to mention that he will be employed by ACE.

Engineer E requests that Engineer A should bring all of his work assignments to a point of completion that will facilitate his making an orderly transfer to other U&I employees and to conclude as many assignments as possible before departing. For the next two weeks before leaving U&I, Engineer A continues to negotiate and prepare recommendations on bids including those that had been submitted by ACE.

Questions:

1. Was it ethical for Engineer A not to volunteer to U&I the information that he would be employed by ACE within two weeks?
2. Was it ethical for Engineer D to entice Engineer A to consider employment with ACE?
3. Was it ethical for Engineer A to interview with a supplier of U&I without first advising U&I of his intent?
4. Was it ethical for the ACE Vice President of Engineering to offer employment to Engineer A without first divulging the risk that the offer might be withdrawn by the ACE CEO?
5. Would it have been ethical for U&I to have interfered in Engineer A's employment change had U&I become aware of the identity of the future employer and ACE's susceptibility to pressure from U&I?

CASE STUDY 4

Public Welfare—Bridge Structure

Case Scenario:

Engineer A, who works for a local government, learned about a critical situation involving a bridge 280 feet long, 30 feet above a stream. This bridge was a concrete deck on wood piles built in the 1950s by the state, and was part of the secondary roadway system given to counties.

In June 2000, Engineer A received a telephone call from the bridge inspector stating this bridge needed to be closed due to the large number of rotten piling. Engineer A had barricades and signs erected within the hour on a Friday afternoon. Residents in the area were required to take a 10-mile detour.

A detailed inspection report prepared, signed, and sealed by a consulting engineering firm, indicated seven pilings required replacement. Within three weeks, Engineer A obtained authorization for the bridge to be replaced. Several state and federal transportation departments needed to complete their reviews and tasks before the funds could be used.

A citizens' rally was held and a petition with approximately 200 signatures asking that the bridge be reopened to limited traffic was presented to the County Commission. Engineer A explained the extent of the damages and the efforts under way to replace the bridge. The County Commission decided not to reopen the bridge.

Preliminary site investigation studies were begun. Environmental, geological, right-of-way, and other studies were also performed. A decision was made to use a design build contract to avoid a lengthy analysis for the pile design.

A nonengineer public works director decided to have a retired bridge inspector, who was not an engineer, examine the bridge, and a decision was made to install two crutch piles under the bridge and to open the bridge with a 5-ton limit. No follow-up inspection was undertaken.

Engineer A observes that traffic is flowing and the movement of the bridge is frightening. Log trucks and tankers cross it on a regular basis. School buses go around it.

Question:

What is Engineer A's ethical obligation under these circumstances?

CASE STUDY 5

Misrepresentation of Education

Case Scenario:

Engineer A is a professional engineer who occasionally provides forensic engineering services as part of the litigation process. As part of a written submission during a legal proceeding, Engineer A indicates that he possesses a degree in electrical engineering and a doctoral degree in electrical engineering. In fact, Engineer A's baccalaureate degree was in engineering technology and his doctoral degree was an honorary degree bestowed upon him by an engineering school.

Engineer B, who knows Engineer A, learns of these misrepresentations in discussions with his colleague, Engineer C, who is serving as an expert witness for the side opposing Engineer A's client. Engineer C is unaware Engineer A is misrepresenting his educational credentials.

Question:

What is Engineer B's ethical obligation under the facts and circumstances of this case?

CASE STUDY 6

Former Consulting Engineer for Utility as Expert Witness for Utility Consumer Complaint Case

Case Scenario:

State Electric Light Company Inc. (SELCI) is a privately owned electric utility company that is part of a group of utility distribution companies. For several years, SELCI's parent company has hired an electrical engineer as a consultant to do several studies for SELCI. This consultant, Engineer B, has recently been asked to testify for Consumer X, a party that has filed a complaint to the State Public Utilities Commission (SPUC) against SELCI.

Question:

Is it ethical for Engineer B to testify for a party that has filed a complaint with the State Public Utilities Commission against SELCI?

CASE STUDY 7

Public Welfare—Duty of Government Engineer

Case Scenario:

Engineer A, an environmental engineer employed by the state environmental protection division, is ordered to draw up a construction permit for a power plant at a manufacturing facility. He is told by a superior to move expeditiously on the permit and "avoid any hang-ups" with respect to technical issues. Engineer A believes the plans as drafted are inadequate to meet the regulation requirements. This is because outside scrubbers to reduce sulfur dioxide emissions are necessary and without them the issuance of the permit would violate certain air pollution standards as mandated under the 1990 Clean Air Act.

His superior believes that plans that involve limestone mixed with coal in a fluidized boiler process would remove 90% of the dioxide and will meet the regulatory requirements. Engineer A contacts the state engineering registration board and is informed, based on the limited information provided to the board, that suspension or revocation of his engineering license was a possibility if he prepared a permit that violated environmental regulations.

Engineer A refused to issue the permit and submitted his findings to his superior. The department authorized the issuance of the permit. The case had received widespread publicity in the news media and is currently being investigated by state authorities.

Questions:

1. Would it have been ethical for Engineer A to withdraw from further work in this case?
2. Would it have been ethical for Engineer A to issue the permit?
3. Was it ethical for Engineer A to refuse to issue the permit?

CASE STUDY 8

Engineer's Duty to Report Data Relating to Research

Case Scenario:

Engineer A is performing graduate research at a major university. As part of the requirement for Engineer A to complete his graduate research and obtain his advanced degree, he is required to develop a research report.

In line with developing the report, Engineer A compiles data pertaining to the subject of his report. The vast majority of the data strongly supports Engineer A's conclusion as well as prior conclusions developed by others. However, a few aspects of the data are at variance and not fully consistent with the conclusions contained in Engineer A's report.

Convinced of the soundness of his report and concerned that inclusion of the ambiguous data will detract from and distort the essential thrust of the report, Engineer A decides to omit references to the ambiguous data in the report.

Question:

Was it unethical for Engineer A to fail to include reference to the ambiguous data in his report?

CASE STUDY 9

Engineer Misstating Professional Achievements on Resume

Case Scenario:

Engineer A is seeking employment with Employer Y. As an employee for Employer X, Engineer A was a staff engineer along with five other staff engineers of equal rank. This team of six was responsible for the design of certain products. While working for Employer X, Engineer A worked with five other engineers in his team and they were credited with the design of a series of patented products.

Engineer A submits his resume to Employer Y and on it implies that he personally was responsible for the design of products that were actually designed through a joint effort of the members of the team.

Question:

Was it ethical for Engineer A to imply on his resume that he was personally responsible for the design of the products that were actually designed through the joint efforts of the members of the design team?

CASE STUDY 10

Duty to Disclose Disciplinary Complaint to Client

Case Scenario:

Engineer A is retained by Client B to perform design services and provide a critical path method (CPM) schedule for a manufacturing facility. Engineer A prepares the plans and specifications and the CPM schedule.

During the rendering of services to Client B on this project, the state board of professional engineers contacts Engineer A regarding an ethics complaint filed against him by Client C relating to services provided on a project similar to the services being performed for Client B. Client C alleges that Engineer A lacked the competence to perform the services in question.

Engineer A does not believe it is necessary to notify Client B of the pending complaint. Later, through another party, Client B learns of the ethics complaint filed against Engineer A and tells Engineer A that he is upset by the allegations and that Engineer A should have brought the matter to Client B's attention.

Question:

Was it unethical for Engineer A to not report to Client B the ethics complaint filed against him by Client C?

CASE STUDY 11

Records Relating to Services to Former Client

Case Scenario:

Several years ago Engineer A, a mechanical engineer, consulted for Company A, a pressure vessel manufacturer, on a specific pressure vessel problem relating to the design of a boiler system. Engineer A's work focused on specific design and manufacturing defects that caused deterioration of the boiler system. Engineer A completed his work and was paid for his services.

Now, Engineer A was retained by Attorney X in a case involving a fatal explosion of a recently designed and manufactured pressure vessel at a facility previously owned by Engineer A's former client, Company A. The facility was sold to Company B seven years before the explosion.

The litigation does not involve any of the issues related to the services Engineer A provided to Company A ten years earlier. The defendant's attorney discovered through Engineer A's deposition and statements relating to his professional experience that Engineer A had worked for Company A on a pressure vessel problem.

Engineer A explains to the defendant's attorney that he is not relying on any of his prior work for Company A in this case. Nevertheless, the defendant's attorney requests that Engineer A provide his files from the previous work performed for Company A.

Question:

Would it be ethical for Engineer A to voluntarily release the files to defense counsel?

CASE STUDY 12

Specifying Equipment of Company Owned by Engineer

Case Scenario:

Engineer A is asked by a firm to prepare specifications for an air compression system. Engineer A made the firm aware that she is the president (and major shareholder) of a company that manufactures and sells air compression systems and that she has no problem with preparing a set of generic specifications. Engineer A also provides the firm with four other manufacturers that prepare air compression systems for bidding purposes, and Engineer A did not include her company as one of the four specified manufacturers.

The firm now wants to meet with Engineer A and a salesman from her company. Engineer A indicated to the firm that it might be a conflict of interest.

Question:

Would it be a conflict of interest for Engineer A to prepare a set of specifications for an air compression system and then have her company manufacture the air compression system under the facts?

CASE STUDY 13

Failure to Disclose Other Business Interest

Case Scenario:

Engineer A, a mining engineer, is retained by a company that owns land on which coal mines are located. Engineer A provides engineering services and surveys to determine the location of coal veins in the mine, assigns coal contractors to the locations in the mine, and performs other engineering services as required.

Engineer A also owns a laboratory that evaluates the quality of coal mined by coal contractors that contract with the coal mine owner. The quality and cost of mining the coal may vary.

Although Engineer A mentioned that he owns a laboratory, Engineer A never informs the coal mine owner about the size and the extent of his laboratory, which is substantial and employs several other engineers and technicians, nor about his clients who are mining the owner's coal.

Question:

Was it ethical for Engineer A to not fully disclose the size and extent of his laboratory and his clients to the coal mine owner?

CASE STUDY 14

Software Design Testing

Case Scenario:

Engineer A is employed by a software company and is involved in designing specialized software for the operations of facilities affecting the public health and safety. The design of this software includes extensive testing by Engineer A. Although tests demonstrate the software is safe to use under existing standards, Engineer A is aware of new draft standards that are about to be released by a standard setting organization—standards that the newly designed software may not meet.

Testing is extremely costly and the company's clients are eager to move forward. The software company wants to satisfy its clients, protect the software company's finances, and protect existing jobs; but the software company management wants to be sure that the software is safe to use.

A series of tests proposed by Engineer A should determine if the use of the software is safe. The tests are costly, will delay the software at least six months, will put the company at a competitive disadvantage, and cost them a significant amount of money. During the testing time, the state public service commission utility rates will rise significantly. The company requests Engineer A's recommendation concerning the need for additional software testing.

Question:

Under the code of ethics, does Engineer A have a professional obligation to inform his company of the reasons for needed additional testing and his recommendations that it be undertaken?

CHAPTER

12 Leadership Skills

Leading the Way to Success

"If your actions inspire others to dream more, learn more, do more and become more, you are a leader."

—John Quincy Adams,
2nd President of the United States of America

LEADERSHIP SCENARIO

"As an engineering student you are constantly faced with technical challenges that the engineering curriculum has prepared for you during your four years at college. You have to deliver your homework on time, study new technical terms, equations, and engineering concepts. It is safe to say that during those four years, the majority of your focus will be on solving technical problems.

One of the important lessons to learn during your undergraduate studies is that technical challenges and innovative solutions to technical problems are not the only tasks that you will be performing once you graduate. You need to also develop nontechnical skills such as leadership to be successful as a student and later in your working career."

(Krenar Komoni, Electrical Engineer and Entrepreneur)

Krenar provided a good foundation for our discussion on leadership with his comments indicating that you need more than technical skills to be a successful engineer. After we show you how to develop leadership skills in this chapter, we will see how Krenar used these skills to enhance his education and to become a successful entrepreneur.

12.1 Introduction

Leadership is an important engineering tool, and learning to be an effective leader in various environments will significantly increase your personal and professional success as an engineer. Leadership will be part of your career when you are a leader and when you work with others who are leaders.

LEARNING OBJECTIVES

By using the information and exercises in this chapter you will be able to:

- Understand why leadership skills are an important part of your skill set.
- Apply basic principles of leadership.
- Take advantage of opportunities to develop leadership skills.
- Work successfully with leaders.
- Apply leadership skills to future career opportunities.

So, what do you think is a good definition of a leader? Thousands of history books have been written about leaders. Typically, these focus on individuals who have been heads of state, generals in a military setting, corporate officers, front-runners in a cause, or outstanding individuals in academia and other areas.

As indicated in the quotation from John Adams, leadership is not limited to the famous of history. In fact, it can be found in all walks of life and in some unique settings. And to the benefit of mankind, engineers will continue to be leaders as they apply important technology to help solve many of the challenges of a fast moving world.

Leaders are people who have vision and the ability to inspire others to accomplish great things. There are two reasons why you will benefit by understanding and applying the principles of leadership. First, when you take advantage of leadership opportunities, this will help you achieve greater success. And second, the complexity of life in a world of accelerated application of new technology has created a need for you to assume greater leadership roles.

The purpose of this chapter is to discuss leadership in a way that you can relate to leadership roles in your future and to show you how you can make this an important part of your engineering skill set. This discussion will include:

- Common principles of leadership
- Your opportunities to develop and improve leadership skills
- How you can practice working with leaders
- How you can apply your experience in leadership roles to your working career
- How you can use leadership skills in your profession and community

12.2 Leadership Principles

"Leaders aren't born, they are made. And they are made just like anything else, through hard work. And that's the price we'll have to pay to achieve that goal, or any goal."

(Vince Lombardi, legendary coach of the Green Bay Packers professional football team)

Although it might appear that some people are natural leaders, we should consider the quotation from one of the most successful football coaches in history, Vince Lombardi. People may have personality traits that are leadership oriented, but they still must develop and enhance their leadership skills just like technical and other skills.

A good leader has vision and the ability to see objectives clearly when others may not understand, agree, or be supportive. Leaders use a systematic approach to achieve their objectives.

Leadership skills need to first be practiced and developed while you are a student. As you take advantage of new and expanded opportunities, you typically become involved in large projects that require leadership skills to inspire and enable others to be actively involved. These examples demonstrate that leadership has a strong focus on people.

Before discussing how various leadership principles are applied in different situations, let's review some of the more common and important skills typically demonstrated by successful leaders, as outlined in Figure 12.1.

To help us gain a better understanding of what these leadership skills really mean, let's expand our discussion and look at typical characteristics of each.

Vision

- Looking beyond the present to see the potential of new opportunities
- Taking reasonable risks and enjoying the adventure of new concepts
- Accepting challenge as opportunity

Innovation

- Thinking "outside the box" and stretching existing concepts
- Doing the possible but also trying to do the impossible at times
- Encouraging and supporting others to be challenged

LEADERSHIP SKILLS

Vision
Innovation
Enthusiasm
Strong Character
Decision Making
Quality Work
Self-Confidence
Communication
Interpersonal Relationships

Figure 12.1
Leadership skills.

Enthusiasm

- Seeing challenges as exciting opportunities
- Expressing joy in life and work-related activities
- Keeping a positive attitude when situations become difficult or unexpected

Strong Character

- Defining principles, communicating them to others, and putting them into action
- Building respect by being honest and truthful
- Using the experience of failure to build character and confidence for the future

Decision Making

- Using facts and intuition to make decisions
- Giving others the opportunity to speak before making final decisions
- Taking action after making decisions

Quality Work

- Working from strength by obtaining information and expanding knowledge
- Paying attention to details but keeping the long-term objectives in focus
- Exceeding expectations by underpromising and overproducing

Self-Confidence

- Building new skills and self-confidence, but remaining realistic about limitations
- Avoiding discouragement when situations become difficult
- Developing a sense of perspective by not overreacting to small problems

Communication

- Practicing frequent and effective verbal and written communication skills
- Listening to ideas and concerns of others
- Making sure others understand what you want

Interpersonal Relationships

- Giving accountability to others and encouraging them to be innovative
- Being sensitive to others and applying the Golden Rule
- Being yourself and setting good examples

12.3 Opportunities to Practice Leadership Skills

Let's see how these leadership skills can be developed while you are a student.

12.3.1 Study Groups

Informal study groups offer many important benefits including opportunities to develop team and leadership skills. You can practice interpersonal and communication skills by encouraging your study group to be innovative when making decisions about how to solve problems. Although participating in study groups may not be the most enjoyable investment of your time, if you demonstrate enthusiasm and contribute to the team discussions, you will be developing leadership skills and be more self-confident in group settings.

12.3.2 Laboratory Groups and Project Teams

Use opportunities to practice leadership when assigned to laboratory and project teams. Observe how different team members relate to your leadership efforts and to the dynamics of group activities. This is also a good opportunity for you to learn how to work with other leaders and be an effective team member.

In team settings, leaders apply vision, are enthusiastic about what can be done, and offer ideas on how to accomplish goals. Team leaders are not afraid to make decisions and help team members produce quality results. You can enhance your leadership skills by being innovative in group discussions and using good communication to encourage others to understand and accept your ideas. So, you can develop many leadership skills in addition to learning how to apply technology when involved in technical projects.

"Engineering education tends to place an inordinate emphasis on developing specialized technical proficiencies, and this contributes to a deficient representation of what it takes to be a complete and successful engineer. In practice, above all your technical prowess, people need to first appreciate that you care. The abilities to both formulate and communicate authentic visions are critical toward motivating teams to solve challenging problems. Purposeful reflective judgment and clean verbal and written skills provide the context through which that vision is transformed into action and the professional engineer into an effective leader."

(Dr. Daniel E. Raible, Electrical Engineer at a Federal Aerospace Research Laboratory)

12.3.3 Professional Organizations

We tend to think of professional organizations as a resource for technical information. This, of course, is one of the major benefits of membership in these organizations. However, your active membership in student professional organizations is also an excellent way to practice and improve leadership and many other nontechnical skills.

Because student activities are planned and coordinated by officers and other leaders, you can observe how they lead volunteers. This is a friendly environment in which you can volunteer to work with leaders. And, learning to work with others is an excellent foundation on which to build your teamwork and leadership skills.

As you gain experience working with leaders, you can then consider opportunities to take responsibility and lead projects. There are many reasons why being involved as a volunteer in student organizations will help you develop leadership skills. For example, it is easy to find opportunities to practice team building and other interpersonal skills when you participate in projects.

"Leaders are good at recognizing people's strengths, and building a team that uses everyone's strengths to work towards a common goal. The best leaders are mentors, and build future leaders who continue guiding the team after the first leader is long gone. Leaders never forget their core values, learn from those they're leading, and work for progress towards the goal, not for recognition. Being involved in IEEE taught me what it means to be a leader. I got involved my freshman year and as I gained skills, I was able to help more. I acquired skills in time management, communications, and problem solving. I gained friends and a professional network; more than I ever imagined. It is when something doesn't work out the way it's supposed to, that you learn what you and your team are really capable of doing and attain the confidence needed to take on any task."

(Sonja Abbey, Past Chair at IEEE Ohio University Student Branch)

When you volunteer, practice some of the basic leadership skills in each project. Start by being enthusiastic and communicate your vision to others by explaining why this would be a benefit to them and to the organization. Build self-confidence by accepting leadership roles, as shown in Figure 12.2, to achieve project objectives.

Figure 12.2
Volunteer leadership roles in professional organizations.

The challenge of convincing other people to volunteer to follow your leadership efforts is excellent training for career leadership roles. However, recognize that this is usually a very big challenge.

The most important thing to remember about volunteers is that they have no obligation to be involved. Volunteer means they are giving their time for the benefit of others without the expectation of compensation. So, their time is a gift to the organization.

> *"You do not lead by hitting people over the head—that's assault, not leadership."*
>
> (Dwight D. Eisenhower, 34th President of the United States of America)

Therefore, as indicated by past president Dwight Eisenhower's comments, leadership of volunteers takes on an important new meaning. In fact, if you are successful in influencing volunteers to follow your ideas, you have learned to use some of the most effective tools for career leadership responsibilities.

A typical comment relating to professional organizations is that members are too busy or just not interested in getting involved. So, how do you motivate them to become involved? Guess what? You can't. But you can often establish an environment in which they will become self-motivated and see the value of their involvement.

To develop an environment in which others will become motivated, use good communication skills such as giving clear directions, asking and listening to others' ideas on how to do the project, and enabling others to have a stake in the results. Then, be sure to give them personal and public recognition for their good work. You can be a successful leader when working with others in situations when you have no direct authority if you follow the principles listed in Figure 12.3.

But, what if you are uncomfortable in leadership roles? Should you still consider getting involved in volunteer work in an organization? Yes, because in this setting you can learn how to work with leaders, and this will also be an important skill in your working career.

12.4 Opportunities to Apply and Enhance Leadership Skills after Graduation

When you start your working career, your leadership opportunities will be limited. You will likely have individual assignments under the direction of your supervisor. Other assignments may include your participation on a project team. Until you gain experience, you typically will be working under the leadership of others.

12.4.1 Preparing for Leadership Roles

So, how can you further develop leadership skills in your early working assignments? You can answer this question by using the same principles discussed for study groups, laboratory and senior project teams, and professional organizations.

SUCCESSFUL LEADERSHIP WITHIN VOLUNTEER GROUPS

- Organize around a purpose
- Select volunteers who are genuinely interested in being involved
- Be actively involved and visible with volunteers
- Make sure others understand what you want
- Be loyal and supportive
- Work to avoid disagreement by clearly establishing what is needed
- Encourage and support others to be challenged
- Support innovation and give due credit to others
- Select initial tasks for small and quick successes
- Don't be easily discouraged when others do not meet your expectations
- Develop a sense of perspective by not over-reacting to small problems
- Listen to ideas and concerns of others
- Criticize ideas and not people
- Give public recognition of appreciation for quality work by others

Figure 12.3
Keys to success in volunteer leadership positions.

Your initial role in student group experiences is usually to observe other leaders and learn how to apply many of your skills to support them in accomplishing desired goals. This is also true for early group assignments in your working career.

Even if you are not the leader, you can be enthusiastic and offer innovative ideas. You can demonstrate self-confidence by making appropriate decisions and providing quality work. You can show good interpersonal skills as you work with group members and communicate effectively. When you work successfully with leaders and peers, you can use this experience to strengthen your team building skills.

12.4.2 Group Work Assignments

As you progress in your career, your responsibilities will grow and will usually include leadership assignments. Initially, these may be small projects for you to lead but are excellent opportunities for you to enhance and apply leadership skills.

As a team leader, you can be innovative and use your vision for the project to encourage other team members to achieve team goals. Your enthusiasm and interpersonal and communication skills will demonstrate that you have the self-confidence to be a successful team leader.

When necessary, a good leader demonstrates strong character skills by making decisions that are best for the project, even if all members do not agree. If leaders show high quality in their contributions to the team, they will set a good example for other members.

In addition to their personal involvement in a project, effective leaders move their focus from "I" can do this to "we" can do this much better. Methods to develop a "we" focus are shown in Figure 12.4. A leader does not do all the work as an individual. An effective leader brings people and resources together for greater results.

Effective leaders learn that having authority within a group does not, by itself, guarantee success. Instead of using authority to direct group members, strong leaders develop the ability to provide information in a convincing way so others become motivated to be supportive and involved.

DEVELOPING A "WE" FOCUS

- Be yourself and set the right example to help others grow professionally
- Under promise and over produce
- Be sensitive to others and apply the Golden Rule
- Analyze strengths as well as weaknesses in others
- Learn to work with different personality traits in others
- Have enough information about what others can do to judge their ideas
- Match people to tasks
- Develop teamwork and include diversity
- Learn how to successfully delegate responsibility
- Confront disagreement with facts
- Learn the value of compromise and use it to achieve greater results
- Encourage and support others to be challenged
- Be actively involved and visible with subordinates
- Practice frequent and effective communication skills
- Give everyone the opportunity to speak before decisions
- Move on after making decisions
- Listen to ideas and concerns of others
- Pay attention to details but keep the long-term objectives in focus

Figure 12.4
Focusing on others.

Leaders establish an environment that encourages others to become motivated.

An effective leader requests information, listens to ideas of others, and uses their ideas when appropriate. Leaders understand the value of ownership, and give credit to others for their ideas to build their team support.

If a leader tries to take credit for work of others, the overall success will be much lower than when all involved feel a part of the results. So, humility and the ability to encourage others through praise and respect are important skills for leaders.

Because the exchange of information is vital to any project or group activity, a leader should possess and use effective communication skills. This includes all four forms of communication—listening, reading, writing, and speaking. Communication should be frequent and clear. It is not limited to within a working group, but should be directed to all persons who have a stake in the results.

Up to this point in our discussion of group situations, we have focused on the leader. While you probably expected that, we should also discuss another part of leadership.

We might assume that most engineers are capable of completing individual assignments. But, not everyone is comfortable or even desires to be a leader in a group setting. If you fit in this category, you might ask, "How does the information in this chapter apply to me?" This is a good question, so let's think about that for a moment.

You can work more effectively with leaders by understanding leadership principles.

In a team or group situation, there should be only one leader. But, everyone in the group needs to work with that leader. One of the best ways to do that is to understand what is involved in being a leader and respond to his or her leadership in a way that will support group success.

This helps the leader to be successful and, even more importantly, supports the success of the project. So, even though you may not be the leader, your ability to understand the principles of leadership and to work effectively with leaders and other members of the group will have a positive impact on your career.

12.4.3 Professional Organizations and Community Activities

Continuing active membership in professional organizations after graduation is one of your best methods of keeping your technical skills up to date. Societies provide a wealth of information on changes in technology. Perhaps even more important, these same organizations offer you opportunities to apply your leadership skills and continue to improve your networking and interpersonal skills.

Your leadership and project management skills can be a great asset to address important professional issues. Professional societies were formed to advance technology and to serve the needs of members of the profession. As a volunteer member of one or more professional groups, you can contribute to the advancement of technology and to help students be better prepared for successful careers.

The world continues to grow more complex because of the many changes resulting from expanding technology applications. In this environment, you can play a major role in helping your community understand and use technology in an efficient way. Your leadership skills are always welcomed by those who have the responsibility to provide community resources.

> ***"Leadership is the practice of translating institutional goals into a plan of action, sharing one's vision with peers, subordinates, and supervisors, and guiding teams to achieving goals. Professional societies provide volunteer opportunities and this is a wonderful platform for individuals at all experience levels, from students to seasoned executives, to develop leadership skills. I have personally been involved in IEEE leadership since becoming a graduate student, and my education, career, and volunteerism have been inextricably joined ever since. The benefits that have accrued to me, in addition to executive leadership training, include several life changing experiences."***
>
> **(Monica Mallini, PE, Electrical and Mechanical Engineer, and Assistant Professorat Northern Virginia Community College)**

Although involvement in community projects requires time investment, the experience gained is helpful to your career as well as to the community. Because you develop special skills as an engineer, you should give serious consideration to being an active leader in your community and sharing those skills and knowledge with students, community leaders, and other important community groups. This experience will bring personal and professional rewards, and is an excellent way for you to contribute to making the world a better place.

A CONVERSATION WITH KRENAR KOMONI

Krenar, how did you develop leadership skills when you were a student?

Answer:

"During my junior year in college I had the opportunity to run for the IEEE student branch president, and through the support of my classmates I was elected that year. This was a great platform where I could learn, improve, and exercise my leadership and communication skills to energize our branch and organize multiple events.

While a senior I was elected for a second term, and in a span of two years we organized more than twenty events, participated in more than ten field trips, and brought nine speakers to talk about their engineering experiences after they had graduated. In addition to IEEE, I was also elected as the president of

Tau Beta Pi Engineering Honor Society—that year our school received the Most Outstanding Chapter Award from the Tau Beta Pi Association."

What nontechnical skills did you improve as a leader?

Answer

"As the IEEE student branch leader, one of the nontechnical skills that I cultivated was the ability to plan. I soon realized that planning is what lets us create a vision and a mission for the organization and people that you are leading. The vision that I created for the IEEE student branch is what guided me, my executive team, and the members of the branch to improve our organization, organize as many events as we did, and generate financial surplus at the end of each year. These skills still remain with me today, and have helped me to start and grow my own businesses.

In addition to planning and creating a mission and a vision for the organization, there are a plethora of other nontechnical skills that one uses when leading a team. The nontechnical skills that I use most frequently are time-management, ability to hold effective meetings, and communication skills. During any engineer's undergraduate experience, I believe that there needs to be emphasis on these skills, especially the communication skills.

The ability to clearly share your thoughts and ideas in front of an audience is the key to any leader's success. Once you create the vision for the organization, communicating that vision to your teammates is crucial. This allows you as a leader to create a good teamwork environment and get everyone to work toward a common goal.

Another nontechnical skill that has helped me tremendously during my undergraduate studies was time management. The discipline that I developed, while exercising my time management skills, has helped me focus on the tasks at hand instead of thinking of the upcoming tasks that only create distractions.

Finally, an important skill that I learned from my professors at my undergraduate school was the skill of conducting effective meetings. Effective meetings need to have an agenda, they need to have an allocated time for particular items in the agenda, and one has to make sure that the agenda is shared with all the participants in a meeting. I used to print-out twenty-to-thirty agendas for every meeting and distribute them to everyone that participated – this was the best way to get everyone engaged on what we were doing, and at the same time get the most involvement from them."

How do you use the leadership skills you developed in school in your career?

Answer

"The leadership opportunity that I had during my undergraduate degree has helped me a great deal in starting my own businesses and becoming an entrepreneur. I firmly believe that my participation in the leadership opportunities that I had as an undergraduate has shaped my character and has prepared me for the uncertainties that come with the entrepreneurship career.

After my bachelor's degree, I was lucky to get hired as a design engineer for a RF wireless engineering company. Working in a startup environment, I had to present my ideas every few weeks and show progress in front of the entire engineering team. There, I realized that presentation and communication skills that I had cultivated during my undergraduate degree were beneficial and have directly helped me enhance my career and have given me the confidence to start my own business."

What advice would you share with students?

Answer

"Compared to other majors, during the pursuit of your engineering degree you will be faced with a plethora of leadership opportunities. There is a great number of engineering organizations that have student branches at your school.

You can easily take the initiative and become the leader of one of these prestigious engineering student organizations such as: IEEE, Tau Beta Pi Engineering Honor Society, Eta Kappa Nu Engineering Honor Society, American Society of Mechanical Engineers (ASME), American Society of Civil Engineers (ASCE), Association for Computing Machinery (ACM), and many other organizations that have branches at your school.

If one of these organizations sparks your interest, then I would encourage you to get involved and volunteer for leadership positions.

You will be a more effective leader and achieve greater career success when you:

- Are open to opportunities.
- Expect challenges.
- Don't let fear of failure stop you from achieving your dreams.
- Follow your passion."

(Krenar Komoni, Electrical Engineer and Entrepreneur)

12.5 Conclusion

The many technological changes taking place in the twenty-first century lead to a growing need for engineers to be leaders. And the good news is, engineers can and do make excellent leaders. In the future, more engineers will rise to high-level management positions. And, engineers who are not managers can become effective leaders in other ways.

Start now to practice and develop leadership skills. This will prepare you for better opportunities in your working career. It will also help you decide if you want to be a leader or, if not, how you can better work with leaders.

Outstanding opportunities for you to develop leadership and self-confidence can be found in your student professional organizations. This is one of your best, and rather inexpensive, "laboratories" in which you can improve leadership skills.

Leadership skills are definitely associated with being an officer in any organization. While assuming the responsibilities of an officer does help develop leadership skills, there are other ways to enhance these skills as well. You can chair a committee or work group within the organization. You can also participate as part of a team or committee and help the leader bring direction and success to the project.

One final thought. If you really do not aspire to be a leader in team and other group situations, you can still be an important part of the role to be played by the engineering profession. Your support of leaders and your honest, ethical, and competent work all help the technical and nontechnical world enjoy the benefits of technology applications. So, in either case, you are an important part of the future of our global society.

END OF CHAPTER REVIEW QUIZ

Select the most appropriate answer to the following statements.

1. The most important reason you should develop leadership skills is
  a. To make a lot of money.
  b. To be more successful when leading or working with leaders in team projects.
  c. That this will guarantee you will become an officer in your company.
  d. That people in your community want you to lead projects.

2. Most leaders develop effective skills associated with
  a. Vision and the ability to direct others.
  b. Interpersonal relationships and communication.
  c. Making important decisions.
  d. All of the above.

3. Most leaders are successful in
  a. Managing their personal lives.
  b. Establishing an environment in which people become motivated.
  c. Making a lot of money and being liked by their peers.
  d. Taking care of many details associated with a project.

4. Leadership skills can be best developed in early group projects by
  a. Observing how the leader uses various skills to obtain group success.
  b. Giving the leader advice to help them direct the team.
  c. Doing your part of the project as assigned.
  d. Thinking about how your leadership style would be better.

5. The leader of a project team is most effective by
  a. Doing most of the work.
  b. Making sure everyone on the team knows he or she is the boss.
  c. Using a "we" attitude and not an "I" one.
  d. Getting the team started and then staying out of most of the discussions.

6. When working with volunteers, leaders can enjoy their cooperation if they
  a. Support innovation and give them credit for their contribution.
  b. Tell people how to do things and then leave them alone.
  c. Give them strong direction and require them to give periodic reports.
  d. Disagree with their ideas if they do not agree with what you think.

7. One of the greatest values of active membership in professional organizations is that
  a. It looks great on your resume.
  b. You can be an officer and tell others what to do.
  c. You can practice leadership in a comfortable and inexpensive environment.
  d. You can listen to technical presentations.

8. The general public needs engineers to take a more active leadership role because
  a. Engineers know how to use technology for the common good of society.
  b. There are too many lawyers in leadership positions.
  c. The Internet should be controlled by technical people.
  d. None of the above.

9. If you do not like to be a leader, you can still function on a team by
  a. Just doing what you are told to do.
  b. Understanding leadership principles and learning how to work with leaders.
  c. Not making trouble when you don't like what the leader does.
  d. Making sure the leader likes you and is your friend.

10. It is important to enhance leadership skills to
  a. Prepare for unexpected opportunities for leadership and career advancement.
  b. Learn how to work with others in a team project.
  c. Enjoy greater career success.
  d. Achieve all of the above.

EXERCISES TO DEVELOP AND ENHANCE YOUR SKILL SET

Exercise 12.1 When assigned by your instructor, respond to the following case study questions regarding, "A CONVERSATION WITH KRENAR KOMONI," which appeared at the end of the chapter:

1. What were the two most important leadership skills used by Krenar to establish an environment for other students to become involved in the IEEE Student Branch activities?
2. What two significant nontechnical skills did Krenar use to be a successful team leader?
3. What is an important key in holding effective meetings?
4. How did leadership experience as a student help to prepare Krenar for success as an entrepreneur?
5. What is one of the greatest opportunities to develop leadership and teamwork skills during the undergraduate education experience?
6. Discuss what is involved in Krenar's recommendations for greater success when you
   - Are open to opportunities.
   - Expect challenges.
   - Don't let fear of failure stop you from achieving your dreams.
   - Follow your passion.
7. What is the most important message for you in this discussion of leadership by Krenar?

Exercise 12.2 When directed by your instructor, use the form in Chapter resource 12.1 to complete a self-evaluation of your leadership characteristics. Participate in a class discussion relating to how your results of this evaluation indicate your leadership skills.

If this is not included in a class session, you can use Chapter resource 12.1 to identify leadership skills. High scores in the 70–100 range indicate you have strong leadership skills. High scores in the 31–69 range show that you have some leadership skills and that you can learn to enhance leadership skills to be more effective. If scores are 30 or lower, you will likely be most effective in team situations when you are not the leader.

Exercise 12.3 Maybe you are unsure whether or not you want to be a leader or if you would be an effective leader. One way to help you discover your leadership qualities is to take part in a group activity. The first part of this assignment is to join or establish a study group.

Assignment—Study Group Leadership

Consider your study group as a team. You can elect one of two options for this exercise.

Option 1: *Volunteer to be a leader* of the group for one month. Practice leadership skills and use the project management process to develop a plan for study activities for that month. Request input from your group members and establish an overall objective, a schedule of meetings, benchmarks to measure success, and how to record results.

Prepare a written study group plan and distribute it to all team members. Involve them in the planning and implementation process so they will have ownership. Work to establish an environment in which all will become motivated to participate and contribute to the overall objectives.

At the end of the designated time to practice this project, discuss how future activities can be continued and perhaps improved. Suggest that other group members take turns as leader of the group in a similar manner as you have demonstrated.

If you select this option, complete your assignment as outlined by your instructor and prepare a written report of your experience using the format in Chapter resource 12.2.

Option 2: *Practice working with the leader* of the study group. Review leadership principles of this chapter and identify how the leader is operating. Think about how you can help the leader do a good job so the study group can accomplish a maximum amount of results in the allocated time.

Although not the leader, you should be comfortable in suggesting ways the study group can be more efficient or more productive. Tell the group about your involvement in this exercise so they can help you have information to prepare your written report.

If you select this option, complete your assignment as outlined by your instructor, and prepare a written report of your experience using the format in Chapter resource 12.2.

END OF CHAPTER RESOURCES

Chapter resource 12.1

LEADERSHIP CHARACTERISTICS SELF-EVALUATION			
	PERCENT OF TIME		
CHARACTERISTIC	0–30	31–69	70–100
VISION			
Recognize Potential Opportunities			
Take Reasonable Risks			
INNOVATION			
Think Outside the Box and Explore New Ideas			
Encourage Others to Stretch			
ENTHUSIASM			
View Challenge as an Exciting Adventure			
Have a Positive Attitude			
STRONG CHARACTER			
Build Respect by Being Honest			
Learn from Failure to Increase Future Success			
DECISION MAKING			
Base Decisions on Facts			
Take Action after Making Decisions			
QUALITY WORK			
Pay Attention to Details			
Exceed Expectations			
CONFIDENCE			
Develop New Skills			
Avoid Discouragement			
COMMUNICATION			
Give Frequent and Effective Messages			
Listen as Well as Speak			
INTERPERSONAL RELATIONSHIPS			
Encourage Others to Be Innovative			
Delegate Responsibility to Others			
Be Sensitive to Needs of Others			
Be Sincere and Set a Good Example			
TOTALS			

Chapter resource 12.2

Study Group Leadership Report

NAME (Your Name) ________________________________

Study group description (group member names, meeting times and location, etc.)

Major objectives

Benchmarks to measure success

Documentation process

Your role in this study group and your evaluation of this assignment

13 Career Management Skills

Career Management—An Incredible Personal Journey

"If you don't know where you are going, any road will take you there. However, you may not like the results. Floating through life without planning is the same as taking a long road trip without a map or sailing the seas without a compass."

—Jim Watson, PE, President, Watson Associates

CAREER MANAGEMENT SCENARIO

"Because your world is dynamic, your first career path might include some detours. If you are dissatisfied with your first job, should you consider making a career change? If so, what can you do to take advantage of new and different opportunities? How can you prepare for a major change in your work and life? How can you be proactive and take control of your future?"

(Kathryn Paine, PE, Mechanical Engineer and Director of Engineering & Inside Sales at Mechanical Products SW, Inc.)

These are excellent questions posed by Kathryn Paine. We will see how she successfully handled career changes in our discussion with her at the end of this chapter. To get us started, let's think about your career and how you can use career management to guide you through the many opportunities and challenges of your engineering experiences.

13.1 Introduction

By starting now to manage your career, you can prepare for future challenges and opportunities and enjoy an incredible journey. This chapter provides practical tools for you to develop and use a dynamic career management plan.

The engineering profession is a culture.

LEARNING OBJECTIVES

By using the information and exercises in this chapter you will be able to

- Understand why personal career management is so important to your success.
- Identify what is important in developing a career plan.
- Develop a professional and effective career plan.
- See why writing and implementing action plans are critical to success.
- Document and evaluate results for future planning.
- Control your career through career management.

Before discussing the details of career planning, let's think about how it relates to our profession. The engineering profession is actually a culture. It includes formal education, lifelong learning, experience, and a focus on helping people by solving problems. These are best accomplished by applying technology creatively, efficiently, and ethically to improve the quality of life.

For many years, industry and government have been seeking engineers who are technically competent. They also want engineers who have nontechnical skills and can work successfully with various levels of management while simultaneously serving as leaders and members of project teams. So, you need more than just technical knowledge and skills.

> ***"While maintaining technical skills is of paramount importance—employability also depends on 'soft' skills and competencies. Time management, verbal and listening skills, written skills, and career management are of vital importance."***
>
> **(Comments from the IEEE Industry 2000 Conference)**

The challenge is to include the development of both technical and nontechnical skills in your career plan to secure better job opportunities. When you have a balance of these skills, you are also better prepared to take advantage of opportunities to advance in your career.

There is another reason to include nontechnical skills in your career plan. Although you may be familiar with technology, those in the nontechnical world do not share your knowledge or experience. Technology may be somewhat confusing to others including co-workers in nonengineering departments and the general public. When people do not understand what you do and why it is important, they may create negative impacts on your career.

To make sure others understand your contributions as an engineer, you need to write and speak clearly, and to use other nontechnical skills to make a positive impression on others. By using the career management concepts in this chapter, which include a balance of developing technical and nontechnical skills, you will be prepared to work in a nontechnical as well as a technical world.

13.2 Your Engineering Career

Your interest in engineering probably started long before you enrolled in engineering classes. You may have taken things apart to see what makes them "tick," and then put them back together just for the joy of accomplishment. You probably did well in mathematics and science in high school. And, based on your interests, you are now on your way to an engineering career.

Your career starts the moment you enter the door as a freshman.

You might think your career will start when you graduate from your engineering program. But your engineering career actually began when you started your freshman year in an engineering college or even before. Your early career decisions include selecting a specific type of engineering and then choosing elective courses that will complement your area of interest. You can maximize your educational experience by starting now to plan your career.

"There is no such thing as engineering job security. Jobs are controlled by others and that typically removes most forms of security. However, engineers can develop career security. So, to be successful, you need to plan and control your career."

(Larry Dwon, PE, Electrical Engineer, retired, American Electric Power Corporation)

This quote, from a very knowledgeable and experienced engineer, Larry Dwon, is based on the more than sixty successful years of his engineering practice, and it is an important message to every engineer today.

When you graduate and start your working career, you will soon realize that all engineering positions include some risk. Technology or economic conditions can rapidly change, resulting in the elimination of jobs. If you work for others, they control your future. Even if you own your own company, your clients and your employees have a major impact on what you can do.

And, this raises an interesting question—when other people impact your work, how can you control your future? The answer is to be proactive and plan your career

short height may prevent him from achieving that part of his vision. Can you picture him in a "dunk" contest?

As indicated by the quotation from Theodore Roosevelt, it is better to start with high expectations and fall a little short, than to never reach for the stars and to settle for much less. So, first think about what is important without limitations. Then, be realistic and consider possible restrictions as you complete your plan.

13.3.2 Professional Vision

Your professional vision is what you want to achieve.

Develop a professional vision by considering what you want as it relates to your work. What is your major professional focus? Is it to be a generalist or a specialist? Do you want to know a little about many areas of technology or as much as possible about only one or two areas?

Other questions to consider relate to your type of work. For example, are you interested in basic research for future product development? Or would you rather develop initial designs by computer modeling? Would you like to use your creativity and apply general concepts to specific products? Or, do you want to test new products and designs? Would you like to be involved in the production of products or in product sales?

In addition to traditional engineering jobs, you can consider other engineering-related career opportunities including management, consulting, entrepreneurship, and academia. Engineers often select these career opportunities after gaining experience in earlier job assignments.

The second area of your professional vision relates to your working environment. Where do you want to locate geographically? Do you want to work for a large or small corporation or do you want to own your own company? Are you interested in a job in government or private sector?

Other considerations relate to what you want to achieve. Is it respect, financial security, authority, expertise, or personal satisfaction of a job well done? Answers to these and the other questions relating to your professional vision provide direction to the development of your career plan.

Figure 13.3 is an example of one professional vision statement. Additional examples are included in Chapter resource 13.3.

My civil engineering focus is water resources. I want to develop efficient methods to use water for the production of electricity, for quenching the thirst of citizens in third-world countries, and for adding nutrients for agriculture. I desire to work in the field as well as in an office. My greatest accomplishment would be to use engineering skills to help others.

Figure 13.3
Student professional vision example.

Although you may find it difficult to define work activities early in your educational experience, you can investigate options now to form your professional vision. Remember, a career plan should be dynamic, so you can always make changes to reflect new ideas and opportunities.

When you have a clear vision, you are more likely to achieve higher levels of success, both personally and professionally. Let's consider the following example and see how it demonstrates why reaching a high level of success is beneficial for you.

Many electric utility companies have made substantial investments to increase the voltage (potential) of their transmission grid from 138 kV to 765 kV or higher. This is done because the amount of electric power that can be sent over a transmission line is directly proportional to the voltage. So, utilities use a higher transmission voltage to deliver more product (energy) at a lower cost and greater value for customers.

This example of increasing transmission voltage or potential can be easily related to your career. There are many benefits in achieving your highest potential accomplishments. When you deliver more results in your career, it is good for you, your family, the company for which you work or own, and the clients who use your product or service.

"My professional vision is to be constantly tackling challenging problems. I enjoy learning, and I want to be able to apply my knowledge to solve new and interesting real-world problems. I prefer dealing at the 'big-picture' level, because there's more scope for trying out various solutions. My career goals are to become technically proficient within my field of engineering, and leverage my business and people skills to be an effective problem trouble-shooter."

(Akshay Kashyap, Electrical Engineer, and Financial Technology Associate at Citadel Investment Group)

13.3.3 Strategy

Using your personal and professional vision statements as a foundation, your next step in your career plan is to develop a strategy to achieve results. As shown in Figure 13.1, this includes setting goals and identifying skill needs. Let's address the first part by focusing on the process and value of setting goals that relate directly to your vision.

Goals are targets of desired achievements.

Set Goals Goals should be specific, measurable, achievable, and relate directly to your visions. Examples in Figure 13.4 demonstrate the difference between weak and strong goal statements. Weak goals are typically too general to be measured.

SETTING GOALS	
WEAK EXAMPLES	STRONG EXAMPLES
Get an "A" in physics	Learn basic principles of physics
Be rich	Learn how to invest for profit
Have extra free time	Develop a time management system
Be healthy	Exercise and control diet
Have a coop job	Use resources to find a coop job
Retire when I am 55 years old	Make a career plan to retire early

Figure 13.4
Weak and strong examples of goals.

As you define goals, review your visions and write goals that will help you achieve what you want. Be realistic, but don't be afraid to stretch a little. It is better to plan for high achievement and reach most of the way than to establish activities that are not challenging.

Three types of goals should be considered and established—long term, midterm, and short term.

Long term goals

- Should focus on the big picture
- Are based on what you want many years in the future
- Should relate to your vision
- May be more general than mid or short term goals

Midterm goals

- Should focus on middle years of your career
- Are easier to establish because you do not need to look as far into the future
- Can be more structured than long term goals
- Should support long term goals and visions

Short term goals

- Should focus on your present situation
- Are much easier to develop
- Can be easier to achieve because the present is more stable and defined
- Should support midterm and long term goals and visions

To expand our goal discussion, let's use Figure 13.5 to develop typical goals to support the personal vision in our example.

RELATIONSHIP OF GOALS TO PERSONAL VISION

Personal vision established in Figure 13.2:

My personal vision is to develop a secure life for me and my future family, and have fun doing it. Although I want a job with a reasonable income, my family is really the most important part of my life. I want to participate in my community by being involved in my local schools.

Related goal examples:

Long term:
- Establish a 401k account, and invest in long term securities
- Plan family trips for pleasure and to see new areas of the world
- Take an active role in the Parent-Teachers Association and schedule field trips for high school students to observe engineers at work

Midterm:
- Establish a savings plan for college expenses for my children
- Manage my time to reduce stress and to plan family activities
- Develop a plan and work with math and science teachers

Short term:
- Use a personal budget to balance financial needs
- Develop an effective time management system
- Establish a mentor program to help engineering students improve math and physics skills

Figure 13.5
Goals related to personal vision.

As you review the examples of goals in Figure 13.5, notice that they are identified as long term, midterm, and short term. Each goal has a direct connection to the personal vision.

A similar approach, shown in Figure 13.6, can be applied to the professional vision.

As in the case of goals for the personal vision, the goals in Figure 13.6 are listed as long term, midterm, and short term. Each goal has a direct connection to the professional vision. Additional examples of goals are included in Chapter resource 13.3.

Identify Skill Needs The next step is to review your present skills and identify skills that you need to develop or improve. Start this process early in your educational experience because you have many choices during these years. Review your skills and identify what is needed to achieve your goals.

Skill needs will help you define what to improve to reach your goals.

RELATIONSHIP OF GOALS TO PROFESSIONAL VISION

Professional vision established in Figure 13.3:

My civil engineering focus is water nsounes. I want to develop efficient methods to use water for the production of electricity, for quenching the thirst of citizens in third-world countries, and for adding nutrients for agriculture. I desire to work in the field as well as in an office. My greatest accomplishment would be to use engineering skills to help others.

Related goal examples:

Long term:
- Attend local and international conferences on water resources
- Coordinate major projects that include design and implementation
- Develop leadership skills in ASCE international activities

Midterm:
- Find co-op positions with both indoor and outdoor assignments
- Obtain a master's degree in civil engineering
- Pass PE exam and practice as professional engineer

Short term:
- Select elective courses that nlate to water nsoUl'ce utilization
- Pass FE exam and prepare to take PE exam
- JoinASCE and be an active member

Figure 13.6
Goals related to professional vision.

Identifying skill needs is an integral part of planning and preparing for potential opportunities. To achieve your vision requires more than just being lucky. People who appear lucky usually are proactive and not reactive, as indicated by Thomas Jefferson. They plan and prepare for future opportunities by developing appropriate skills.

As we discussed earlier, skill needs are both technical and nontechnical. When establishing your career plan, identify skills in both areas. During the undergraduate education process, you have options for selecting your major area of study and elective courses. Select courses that help you develop the technical skills that you identified in your career plan.

> *"I'm a great believer in luck, and I find the harder I work the more I have of it."*
> (Thomas Jefferson, 3rd President of the United States of America)

Technical Skill Needs

To achieve the maximum benefit from each course, start by asking, what technical skills can I improve by actively participating in this class? How will this support my strategy and visions? What extra things can I do to maximize the value I receive from this course?

Prepare for class by reading the textbook and writing notes. Use notes to determine what information you need from class so you can ask appropriate questions. This approach will increase your technical knowledge and support your strategy of skill development. Asking questions in class is also a good way to build self-confidence and to learn how to communicate in a group environment.

Consider other technical skills that are more general but still relate directly to your goals. For example, you can use software for a time management system or Internet tools for financial investments. If you are involved in a co-op program or other work-related activities, look for opportunities to develop technical skills that help you be more productive and useful to your employer.

If one of your midterm goals is to become a registered, professional engineer, which we highly recommend, then you should identify specific skills to help you prepare for the Fundamentals of Engineering (FE) exam and PE exam. The opportunities to develop many of these skills will appear well before the date of the exams.

To achieve greater career success, we also recommend that you include earning a master's degree in engineering in your midterm goals. Then, use your present opportunities in laboratory classes to improve technical and nontechnical skills to prepare for future research projects associated with graduate degrees.

Nontechnical Skill Needs

This introduces the second type of skill needs—nontechnical. You will find many opportunities to improve nontechnical skills in your undergraduate education experience. You can improve your communication skills when you ask questions in classes and participate in study groups. You can improve your interpersonal skills by learning to effectively interact with people who are from diverse backgrounds and who may have different ideas.

> ***"As engineers, one of the motivations to which our human mechanism responds to is to correctly communicate technical information—no matter what the culture may be. Today, one of the most basic of human needs—verbal communication—remains vital to success, and the successful engineer should have a working knowledge of various languages, and an understanding of the world's cultures. A thirty-eight year career with a world leading hot and cold forging machinery manufacturer taught me that an engineer's success depends upon a basic knowledge of the customers' language and culture."***
>
> **(Gary Stroup, Mechanical Engineer)**

The classroom is an excellent place to apply ethics, another important nontechnical skill. You are being ethical by studying for exams and not using copies of previous exams (unless this is suggested by the instructor as a study aid). If you use information prepared by others, give proper credit. Cheating is inappropriate and very unprofessional.

Professional organizations offer another great way for you to develop nontechnical skills. These inexpensive "laboratories" offer outstanding opportunities to practice many nontechnical skills in a comfortable environment. When you volunteer to participate in activities, organizations will welcome your involvement.

In addition to listing membership in organizations on your resume, become an active member. This will give you many opportunities to improve important skills such as written and oral communications, teamwork, project management, time management, and ethics. Then, expand your resume with a list of activities to show how you were involved in professional organizations. This often is one of the most important discussion items in job interviews.

Review your present nontechnical skills. Are you comfortable working with other people? Can you write clearly and concisely? Are you confident to stand before an audience and give a presentation? Have you developed effective leadership skills, or do you have experience in project management? Are you a good team member? If the answer to any of these questions is no, then you can add these skill needs to your career plan.

Although we tend to focus on technical skill needs, in reality, nontechnical skills have a greater impact on career success. So, identify important nontechnical skills that you need and target them for short term, midterm, and long term enhancement.

Typical skill needs for goals in the personal vision example are shown in Figure 13.7. Additional skill needs, listed in Figure 13.8, will help reach professional vision goals.

The examples in Figures 13.7 and 13.8 show how to connect skill needs with goals that relate to your personal and professional visions. Additional examples of skill needs are shown in Chapter resource 13.3. Keep these examples in mind as you define your skill needs and make sure they have a direct connection to your goals and visions.

13.3.4 Action Plan

So far we have discussed how you can define what you want in life, what you would like to achieve, and how you set goals and identify skill needs. This is a good start, but it is not a complete plan. As Figure 13.1 indicates, there is more to be done, and the next step is to design an action plan to make your strategy become reality.

> Your action plan brings everything together and includes what, when, and how you plan to achieve results.

An action plan provides details of what, when, and how you control your career. It includes specific activities that relate directly to your strategy. Without an effective action plan, you may not achieve your visions. To be effective, design your action plan to be a road map for the journey you selected when you developed your vision statements.

Start with short term goals and skills that can be developed in the next few weeks and months. Outline activities and define details of what to do, and put target dates

SKILL NEEDS FOR ESTABLISHED GOALS

Goals established in Figure 13.5 relating to personal vision:

Long term goals:
- Establish a 401k account, and invest in long-term securities.
- Plan family trips for pleasure and to see new areas of the world
- Take an active role in the Parent-Teachers Association and schedule field trips for high school students to observe engineers at work

Midterm goals:
- Establish a savings plan for college expenses for my children
- Manage my time to reduce stress and to plan family activities
- Develop a plan and work with math and science teachers

Short term goals:
- Establish a personal budget
- Develop an effective time management system
- Establish a mentor program to help engineering students improve math and physics skills

Skill needs examples:

Technical skills
- Gain experience using Internet tools for investments
- Learn how to use an electronic time management system
- Expand math and physics knowledge

Nontechnical skills
- Learn how to balance expenses and savings
- Learn how to become more organized
- Learn how to work as part of a team

Figure 13.7
Skill needs for personal vision goals.

on each activity. Your action plans should be achievable and measurable and should include a process to record results.

Break up long term activities into smaller steps that you can complete in a reasonable amount of time. Then, you can pat yourself on the back after each successful activity. This may sound insignificant, but it helps you maintain interest in your career management process.

Good-poor examples of action plan items

Good: Schedule 8 PM–9 PM on Monday to read Chapter 3 of the textbook and solve homework problems in Sections 3, 4, and 5 for the statics class.
Poor: Study for the statics class. (Should be more specific and measurable)
Good: Join a study group and spend at least 1 hour three times per week with the group to solve homework problems in calculus.
Poor: Get a good grade in calculus class. (What is a "good" grade and what will you do to achieve it?)

SKILL NEEDS FOR ESTABLISHED GOALS

Goals established in Figure 13.6 relating to professional vision:

Long term goals:
- ◆ Attend local and international conferences on water resources
- ◆ Coordinate major projects that include design and implementation
- ◆ Develop leadership skills in international ASCE volunteer activities

Midterm goals:
- ◆ Find co-op positions with both indoor and outdoor assignments
- ◆ Obtain a master's degree in civil engineering
- ◆ Pass PE exam and practice as professional engineer

Short term goals:
- ◆ Select elective courses that relate to water resource utilization
- ◆ Pass FE exam and prepare to take PE exam
- ◆ Join ASCE and be an active member

Skill needs examples:

Technical skills
- ➢ Expand knowledge of fundamentals of fluid mechanics
- ➢ Increase basic understanding of hydroelectric generation systems
- ➢ Stay current with changes in water management technology

Nontechnical skills
- ➢ Develop methods to obtain the maximum benefit from classes
- ➢ Develop leadership skills
- ➢ Learn to structure and deliver professional presentations

Figure 13.8
Skill needs for professional vision goals.

When developing action plans, it is important to understand the difference between an action plan item and a goal. A goal is something you want to obtain or achieve and an action plan item is what you will do to make that happen.

We can demonstrate the relationship between a goal and action plans to achieve that goal by using an example from one short term goal established for the professional vision. This is shown in Figure 13.9.

An efficient way to develop action plan items is to start with short term goals. Action plans for these are typically easier to establish because you will be implementing them first. Decide what you can do, how you can do it, and when you need to do it to reach your first short term goal. Continue with other short term goals. Then, move to midterm goals and complete the process with long term goals.

After determining action plans for goals, use a similar approach to define action plans to develop new skills identified in your list of skill needs. Examples of action plans for one technical skill and one nontechnical skill associated with the professional vision are shown in Figure 13.10.

ACTION PLAN FOR ONE SAMPLE GOAL

Short term goal to achieve professional vision in Figure 13.8:

Short term goal:
- ◆ Select elective courses that relate to water resource utilization

Sample action plans for this goal:

Action plan items:
- ➢ Meet with advisor to review course options
- ➢ Prepare schedule to include classes associated with water resources
- ➢ Focus on gaining information and improving skills in each course

Figure 13.9
Sample action plan items for goals.

The examples in Figures 13.9 and 13.10 illustrate the difference between action plan items and goals and skill needs. These action plans are specific, measurable, and have a direct relationship with each goal or skill need. Additional examples are shown in Chapter resource 13.3.

Let's summarize the process used to create action plans based on your strategy. Action plans are developed after you establish what you want to do by designing the details of how to do it, as shown in Figure 13.11.

You have a variety of sources of information and activities to help you decide what to include in your action plans. Many of these are associated with classes. You can expand your options by using other resources outside of your classes. If, for example, you need to improve your math skills, join a math study group or use a tutor. If your writing skills are weak, visit the university writing center and request help from their staff.

We associate technical skill development with classes and labs. However, when preparing your action plan, you can also include the development of nontechnical skills in these classes. A good opportunity to build teamwork and interpersonal skills is available when you work with your laboratory team. Other opportunities to enhance nontechnical skills are in classroom exercises relating to ABET accreditation, which we discussed in the chapter on developing your skill set.

"Try out the opportunities your school has to offer. Do an internship. Join a technical student society (preferably the ones that design real projects or participate in professional competitions). Attend corporate recruiting events and talk to the students who interned with the company. Take an undergrad research course with a professor. The most interesting stuff happens outside the classroom; one of the best things you can do for yourself is to find those opportunities and figure out what they mean to you."

(Akshay Kashyap, Electrical Engineer, and Financial Technology Associate at Citadel Investment Group)

ACTION PLAN EXAMPLES FOR SKILL NEEDS

Technical skill to achieve professional vision in Figure 13.8:

Technical skill
- ◆ Expand knowledge of fundamentals of fluid mechanics

Sample action plans for this skill need:

- ➢ Select and complete fluid mechanics and dynamics courses
- ➢ Attend conferences, seminars, and presentations on these topics
- ➢ Join students in study group and work homework problems
- ➢ Select senior design project that relates to these topics

Nontechnical skill to achieve professional vision in Figure 13.8:

Nontechnical skill
- ◆ Develop methods to obtain the maximum benefit from classes

Sample action plans for this skill need:

- ➢ Talk to the instructor to establish a personal network
- ➢ Make a list of what I can learn in this class relating to my vision
- ➢ Establish time schedule and read textbook before each class session
- ➢ Write notes as I read and identify information I do not understand
- ➢ Listen carefully to the class lecture and take effective class notes
- ➢ Use my text notes to ask appropriate questions for clarification
- ➢ Add instructor's answers to questions to class notes

Figure 13.10
Sample action plans for skill development.

We identified how you can improve nontechnical skills by active involvement in professional student organizations. In addition to improving your communication, interpersonal, project management, and teamwork skills, volunteer activities help build self-confidence and add important experiences to your resume. So, think about how you can use these organizations when you develop your action plan.

13.3.5 Action Plan Implementation

"Genius is one percent inspiration and ninety-nine percent perspiration."

(Thomas Edison, Famous American inventor)

PROCESS LEADING TO DEVELOPMENT OF ACTION PLAN

<u>ESTABLISH WHAT TO DO</u>

Strategy to achieve personal and professional visions

- Set goals
  - Long term
  - Midterm
  - Short term
- Determine skill needs
  - Technical
  - Nontechnical

<u>DESIGN DETAILS OF HOW TO DO IT</u>

Action plans to achieve strategy

- Action plan items for goals
  - Short term
  - Midterm
  - Long term
- Action plan items for skill development
  - Technical
  - Nontechnical

Figure 13.11
Action plan development summary.

Up to this point we have reviewed sample personal and professional visions, developed a strategy that includes some goals and skill needs, and combined some of this into an action plan. These are important steps but we are not finished.

Action plans are like the story of three frogs sitting on a log in the middle of a pond. Their names are Bud, Wise, and Ur. A fly landed on a lily pad. Ur did not have an action plan, and so did not know how to catch the fly.

Bud had an action plan and decided to jump off the log to catch the fly. So, how many frogs were left on the log? The logical answer would be two. Sorry, but that answer is incorrect. Bud *decided* to jump, but he did *not implement* his action plan and jump!

Wise, as you might expect from his name, was much more successful. He had an action plan and when he saw the fly he decided to implement his action plan by jumping off the log and eating the fly. The moral of this story is—be wise, have an action plan for potential opportunities, and implement the plan when opportunities arrive.

Action plans are useless without implementation.

The most useless action plan is the one not implemented! Implementing an action plan requires your time investment. When and how you schedule activities has a major impact on how successfully you can complete them on time. So, use time-management to plan, schedule, and complete activities.

13.3.6 Documentation and Analysis of Results

Three additional important steps are needed to complete your plan—documenting results, evaluating results, and fine-tuning your plan. In this section, we will discuss the value of documentation and how recorded results can be used to enhance future success.

Documentation Your action plan should identify the process you will use to document results of the goals you achieve and your success in improving skills. As demonstrated by the student in Figure 13.12, you can establish your documentation process

Figure 13.12
Documentation of career plan results.

now and start recording your accomplishments in classes, laboratories, professional organization projects, and jobs.

It is best to record results soon after you complete each major part of your action plan so you can remember details. Documentation is an important resource for career management because it provides the basis for evaluating results and is a record of accomplishments for future reference.

Documentation of activities relating to work experiences is especially important and useful. If you are fortunate to be involved in co-op or intern positions while in school, experiences gained in these jobs will be some of the best examples of your skill development and accomplishments. Add these to your portfolio, and you will have a more effective story to tell future employers.

Documentation becomes even more important in your future working career. You will probably have many different work positions and may work for many different companies in your career. The best way to demonstrate your skills to a future employer is to use the record of what you achieved in past positions.

Evaluation Your action plan should also include a systematic process of analyzing results. By evaluating what you did, you can determine if you reached goals and developed desired skills that are defined in your strategy to accomplish your visions. This provides the foundation for career plan revisions and helps to determine areas that need additional refinement or changes.

As you evaluate results, you will see that many of your goals were successfully met. Enjoying success is good, but you can learn even more from failure.

In developing your career plan, include some reasonable risks. The key word is reasonable. Then, don't be afraid to fail. You will fail from time to time like everyone else. But remember, failure is not a bad thing, as long as it is not too severe.

So, don't be discouraged when some results fall short of your expectations. Failure is a wonderful teacher that can lead you to greater success. Most successful people fail part of the time, and almost all good products experienced some level of failure in their development process.

By the way, failure is not in falling down—failure is not in getting back up after falling. Review the process that led to failure and learn how you can use this experience to enhance your career. Look back to learn from the past, and then focus more on the future and how your experiences can make you more successful.

Fine-Tune Career Plan This brings our discussion to the last part of the career plan process. Make appropriate changes in future planning based on the results of your action plan implementation. Notice how Figure 13.1 shows a closed loop between evaluation and your visions. You should fine-tune your visions to take advantage of changing conditions and new opportunities.

Even if your vision is still valid, you may want to revise your strategy and subsequent action plans. Many new professional opportunities require new or enhanced skills. You will be more successful in taking advantage of unexpected opportunities when you are proactive and develop skills for possible situations before they appear.

A CONVERSATION WITH KATHRYN PAINE

Kathryn, you raised some very interesting questions at the start of this chapter. How have you managed your career?

Answer:

"Career management and success is typically a lifelong pursuit. I have had three major experiences where I decided to shift the focus of my career.

The first was a major change in career. I had a degree in accounting and came to the conclusion I needed something more challenging. Having lived with an electrical engineer for 10 years, I decided turn-about was fair play. I went back to school and obtained a mechanical engineering degree.

Upon graduating with the engineering degree, I had hoped to pursue work in the field of HVAC. The job market was difficult and I ended up estimating for a tool and die company in NE Ohio. It was still interesting to see how that company functioned and gave me background for a future shift in my career within engineering. Never underestimate the opportunity to learn from what may appear to be a setback.

It did not take long to explore avenues in the HVAC arena and shift my career into consulting engineering. After nearly 5 years with a small firm designing HVAC/plumbing systems for commercial buildings, the opportunity presented itself for the spouse to make a move from the Midwest to the Southwest. At the same time, I had decided I needed to narrow my focus with respect to building systems engineering.

After settling in the Southwest, I managed to secure a position with a manufacturer's representative for HVAC equipment. This third shift was indeed a much more narrow focus by concentrating on particular types of equipment with particular factories. Sales engineering as a profession is perhaps not as technically challenging, but certainly beneficial in many other respects."

How did you decide to make career path changes?

Answer:

"All three major decisions were opportunities that I initiated. In each case, I performed research and took action required to accomplish the goal. I explored college programs available and made various changes in employment for 5 years to accommodate engineering college coursework. To find a position in HVAC consulting, I joined the local chapter of the associated professional organization, ASHRAE (American Society of Heating, Refrigerating and Air-Conditioning Engineers), having been a student member while in college.

It's never too early to start building a professional network. As it happened, it was at a local chapter meeting where I first met my future employer in the consulting world. To secure employment in a new geographic location, I used the relationships I had built while in the consulting business to make contacts with similar professionals in my new town."

What nontechnical skills did you use in managing your career?

Answer:

"Nontechnical skills required for any career management involve research, study, good writing skills as well as people skills and self-promotion. It also requires nontechnical, nonengineering tasks such as self-examination and soul searching in an attempt to learn what you think you desire and recognize the areas where you excel. You also need the ability to revisit and reexamine those same subjects throughout your career in case some adjustment is required.

In all three situations, the career change was a positive experience, filled with the benefits of learning new skills, making new professional contacts, and progressively increased monetary rewards. Each prior experience benefited the succeeding career position."

What advice would you share with students?

Answer:

"A majority of technically inclined individuals such as engineering students tend to view the world in terms of black or white issues, see questions as having yes or no answers, and believe everything should be very clear-cut and well-defined. Math and science have those tendencies.

However, I am of the opinion that you will enjoy greater success in your professional and personal life if you can learn to expand your horizons, view and consider the larger picture as well as maintain an open mind to consider many options. It's important to remember that every experience, whether it's viewed at the time (or even in retrospect) as good or bad, is a learning experience."

(Kathryn Paine, PE, Mechanical Engineer and Director of Engineering & Inside Sales at Mechanical Products SW, Inc.)

13.4 Conclusion

As Kathryn has demonstrated with her personal example of career management, successful careers are the result of taking a personal, active role in planning, controlling, evaluating, and revising what we want to accomplish and then taking steps to be prepared for unexpected opportunities well before they happen.

Although you cannot prevent changes caused by business or other economic situations, you can initiate desired changes if you prepare for new opportunities as described by Kathryn. Do this by reviewing your career status, by setting goals and developing skills, and by being prepared to initiate career changes when the timing is right. In other words, don't knock on that door of opportunity until you have developed the proper skills to be successful in a new career position.

So, who is responsible for your career management? Look in a mirror. When should you start planning and controlling your career? Now! When can you relax and stop controlling your career? Perhaps never, but certainly not before you have achieved your professional vision or retire. Even then, you need to focus on your personal vision for the balance of your life.

END OF CHAPTER REVIEW QUIZ

Select the most appropriate answer to the following statements.

1. The most important thing you can do to achieve greater career success is
   a. Work hard in school and earn a GPA of 3.8 or higher.
   b. Join a professional society and participate in activities.
   c. Balance technical skill development with many nontechnical skills.
   d. Interview for new job openings as soon as they become available.
2. The first step in preparing a career plan is to
   a. Write short-term goals because these can be accomplished quickly.
   b. Develop an action plan so you can start using the process early.
   c. Determine how you will document and evaluate results.
   d. Write personal and professional vision statements.
3. The most important consideration when preparing a personal vision is
   a. What you want to accomplish.
   b. To first decide what you want personally, and then determine the role that your family and community will play in your life.
   c. To think about what you want your life to be like when you are 60 years old.
   d. Ask your parents and friends what you should do in your life.
4. Professional visions statements are most effective when they include
   a. What type of work you want, where you would like to live and work, and what you would like to accomplish.
   b. A lifelong learning plan.
   c. A statement of what kind of work you would like to do and if you want to own your own company or work for a corporation.
   d. Details of work activities and how you will be able to retire at an early age.
5. To be most effective, goals should be developed to
   a. Make sure you do not waste time on unimportant tasks.
   b. Help you achieve your personal and professional visions.
   c. Help you learn new technical and nontechnical skills.
   d. Give you information for discussion with your supervisor so you are working on the correct assignments.
6. When preparing a list of skill needs, you should
   a. Select skills that support your goals and help achieve professional and personal visions.
   b. List skills that you did not learn in previous classes but can be obtained in elective courses during your undergraduate experience.
   c. Identify skills that you feel will be easy so you can enjoy learning them.
   d. Learn skills that will prepare you to own your own company.

7. The most important purpose of an action plan is to
   a. Help you remember your goals and skill needs.
   b. Provide details of your professional vision.
   c. Structure your time-management plan to ensure you complete tasks effectively.
   d. Identify how, when, where, and what you will do to achieve your goals and enhance your skills to realize your personal and professional visions.
8. The greatest value of documenting results is to
   a. Provide a database of information for review and to use in fine-tuning future goals and skill need statements,
   b. Keep a record of your accomplishments to use when you ask for a raise at work or to show your instructor how much you did in class.
   c. Have as a resource in case you would like to write an autobiography.
   d. Find ways to reduce work when using future versions of your career plan.
9. To achieve success in career planning, the most important step is to
   a. Define clear personal and professional visions.
   b. Select a strategy of appropriate goals and skill needs.
   c. Develop, implement, and document an action plan that will help to achieve your skill needs, goals, and visions.
   d. Accomplish all of the above.
10. The greatest value of career planning is to
   a. Achieve a good grade in this class.
   b. Be in control of your life and career, and be prepared to take advantage of unexpected opportunities.
   c. Be organized so you do not waste time.
   d. Accomplish all of the above.

EXERCISES TO DEVELOP AND ENHANCE YOUR SKILL SET

Exercise 13.1 Vision statements

The exercises that follow are designed to help you develop a successful career plan. Complete the following statements as a foundation for developing your visions:

My definition of personal success is ______________________________

____

____

In relation to my career, my family is ______________________________

____

My desired role in my community is ______________________________

____

In my personal life, I would like to be remembered by ______________________________

____

____

My definition of professional success is ______________________________

____

____

My first choice for location to live would be ______________________________

____

The work environment most desirable for me would be ______________________________

____

In my professional life, I would like to be remembered by ______________________________

____

____

Exercise 13.2 Career plan assignment

The following is designed to help you develop an initial career plan.

STEP 1

Use the Career Plan Format in Chapter resource 13.1 to develop an initial career plan.

Review your Career Plan Exercise 13.1 results. Develop your personal vision in paragraph format. Start with what is important to you personally. Then, add a paragraph relating to your family and a paragraph relating to your community.

Review the sample professional visions and write three paragraphs relating to three areas of your professional vision: initial career focus, desired working environment, and overall achievements.

STEP 2

Visualize your life when you are 55 years old. Then, consider what could make that vision come true. Don't be too detailed, but select one general goal that supports the vision of your life at age 55. Express this as a bullet point under long-term goals.

Add bullet points for two midterm goals for the next 5 to 10 years. Repeat the process with bullets for three short-term goals for the next 1 to 5 years.

Review your visions and goals, and select at least two technical skills that will help you achieve them. Add these as bullets under Technical Skills.

Repeat the process for least two nontechnical skills as bullets under Nontechnical Skills.

STEP 3

Develop your action plan by reviewing the first short-term goal. Write one or more action plan bullets to accomplish that goal. Continue this process for each of the other subtopics, ending with an action plan for your Nontechnical Skill Needs. Make sure your action plan items are measurable and that they will help you accomplish your goals, skill needs, and visions.

Complete your action plan by identifying how you will document results. This should be a structured plan that includes periodic documentation and one that can easily be used to review results and to revise future career management plans. Save the file with title of "Your name Career Plan YYMMDD" (year, month, day).

STEP 4

Hand in a printed copy of your career management plan at the start of class on the day established by your instructor and provide an electronic copy to your instructor that day.

STEP 5

Review evaluation and make appropriate revisions to your career plan. For maximum benefits, keep your career management plan dynamic with appropriate changes.

END OF CHAPTER RESOURCES

Chapter resource 13.1

Career plan format

CAREER PLAN

Prepared by ____________(Your Name)____________

Personal Vision

(Three paragraphs relating to you, your family, and your local/professional communities)

Professional Vision

(Three paragraphs relating to your initial career focus, desired working environment, and desired achievements)

Strategy

Goals

Long term

✓

Midterm

✓

Short term

✓

Skill Needs

Technical

✓

Nontechnical

✓

Action Plan

To Reach Short-Term Goals

✓

To Reach Midterm Goals

✓

To Reach Long-Term Goals

✓

To Enhance Technical Skills

✓

To Enhance Nontechnical Skills

✓

Documentation Plan

(Three paragraphs to discuss: [1] how you will document results; [2] how you will evaluate results; [3] how you plan to keep your career plan dynamic and effective.)

Chapter resource 13.2

Outstanding Career Plan Example

by James Sadey

Personal Vision

When I think about my future I never worry about being financially stable, I think about having a family and a wonderful, fulfilling life. This tells me that I will always put my family before my profession. I do not want to be the father who is at work seven days a week and unable to be a part of my children's personal activities. My parents always made time for my interests and that is exactly what I plan to do. By correctly planning my career, I can be the loving father who balances out his profession with his personal life.

Family, in my mind, is the most important part of a person's life. They are the closest people to you and at the end of the day they are all you need. People have wants that they can live without, but living without a family seems like an impossible feat to me. I was raised in an environment where family meant everything. Wants, hobbies, and friends were all a distant second to spending time with each other.

This is what my mindset is when it comes to having my own family someday. I want to be a role model for my children and a loving husband as well. I cannot wait for the opportunity to be the person who my family needs, looks up to, and loves unconditionally. Having a career that I enjoy will satisfy my professional needs, allow me to provide for my family, and empower me to be the person I want to be.

When it comes to my community I have a very simple plan: to display more positive actions than negative. I want a safe environment to raise my family and that might not be an option unless I become an active participant. When it comes to my professional community, I feel the same way. I would love to be involved by displaying positive and ethical practices. I believe by making positive contributions the end result will be constructive.

Professional Vision

My initial career focus is a vague one at this point in time because I have yet to explore the countless areas of electrical engineering. I know for a fact that my passion is in the electrical engineering field but I have not yet decided specifically where I am headed. I am currently interested in designing energy sources because I have always been fascinated with creating new things. The energy field is a popular one at the moment and I am interested in the endless possibilities that can arise in this area.

My desired work environment is outside in the fresh air. I am not someone who likes to be stuck inside for extended periods of time and cannot sit still for very long. I feel like I must always be doing something and being able to work outside would be very enjoyable for me. That being said, if I am focused on something such as building or creating a new idea I can spend hours or maybe days without leaving my work area. Creating things excites me and while being inside is almost a necessity when it comes to our profession, I would love the opportunity to be an active, outdoor electrical engineer.

When thinking about what I would like to achieve in my career, a few things come to mind. Most importantly, I want to learn all that I can about electrical engineering. Obtaining more knowledge will not only help me professionally, but it is actually more of a personal goal. Life is short and I would like to learn as much as possible in the time that I have.

I also have a specific goal that I would like to achieve that I have not mentioned before: My father and brother are both very knowledgeable when dealing with electrical systems and I hope that one day we can all start our own business or company. Working with family in a field that I love would be a dream come true. I dream of being hired by several different companies or industries where we are the people who can provide them a service. Being hired to design electrical systems by possibly a zoo or a company that is outside would be the ultimate high for me because it is a combination of all of my favorite things. I hope one day I can make this dream a reality.

Goals

Long term

- Have a job outside designing electrical systems.
- Own an engineering company with my father and brother.

- Design and create new energy sources.
- Obtain as much knowledge in the electrical engineering field as possible; read theories as well as new technology that is being produced.
- Raise and spend time with my family in a respectable community.

Midterm

- Graduate college with at least a master's degree in electrical engineering.
- Have a few years of experience working at local engineering companies.
- Start saving money toward a car and possibly a house.
- Further knowledge in electrical systems and look into job opportunities in an outdoor environment.
- Create a daily exercise habit to keep health an important factor in my life.

Short term

- Retain a 4.0 grade point average this semester as well as the spring semester.
- Get a summer internship at a local engineering company.
- Join and become an active member in IEEE.
- Create a study group rather than studying alone.
- Ask more questions in class and during office hours.
- Continue to get exercise on a daily basis.

Skill Needs

Technical

- To actually understand and learn new material, rather than just remember it.
- To transition all of my knowledge into the work field, where I can physically apply it.

Nontechnical

- Learn how to trust others and not feel the need to correct them.
- Try to eliminate the need to do all of the work myself.
- Learn how to manage my time efficiently and plan ahead so I can finish work with enough time to enjoy other activities.

Action Plan

To reach short term goals

- Continue to ask daily questions in class to further understand the material.
- Plan ahead to study at least an hour per day for each class outside of school with a study group of friends.
- Read theories behind the material to understand why and how things work.
- Create a solid resume and apply to several companies to obtain a summer internship.
- Participate in announced IEEE activities whenever time is allowed.
- Wake up an hour earlier each day to make time for a run around the neighborhood.

To reach midterm goals

- Integrate master's courses into my senior year so I can graduate with a master's degree in 2015.
- Continually work summer internships during my years at college to further my experience in the engineering field.
- Start saving at least 50% of my paycheck while living at home.
- Travel to companies around the area and ask about possible job openings in the future, and what I can do to increase my chances of obtaining a job from them.
- Work at local engineering firms until I gain the expertise of running my own business.
- Workout for an hour when I come home from work and make time for recreational activities on the weekend.

To reach long term goals

- Research all of the crucial aspects of running a company and start looking for a potential building location.
- Start spreading the word about my company to local businesses and on the East Coast of the country for possible job opportunities.
- Start a family and find an appropriate house and neighborhood to live in.
- Start a college fund for my children and save enough money to fund their expenses.
- Create a weekly planner to figure out an appropriate balance to spend time at work and with my family.

To enhance technical skills

- Instead of doing numerical problems, actually read the theory behind what I am computing and how it works.
- Find online sources about the material I am reading to see another perspective on the material.
- Build objects at home or at work that apply the material in a hands-on fashion.

To enhance nontechnical skills

- Get to know people on a more personal note to build trust between each other.
- Start working at least twice a week with a study group to build friendship together.
- Complete homework as soon as it is assigned and complete projects in the same manner.

Documentation Plan

To document the results of my career plan I have decided the best way to do this is once every week, electronically. Documenting my results every day would be too excessive and it might be impossible to remember details if documentation occurred only once per month. If I set an hour or so every Sunday to document the results of the past week I believe I can properly set myself up for evaluation.

Evaluating my results is an action that I will complete every three months. I should have a good idea of what needs to be completed each year, but to make sure I am on the right track, I will compare my documentation to my original action plan in this continuous interval.

Nothing drastic can happen over three months, to the point that my career plan can no longer be related to, but it is a sufficient amount of time that I can make minor changes as I move along in my career.

My career plan will always be changing. There is no way that I have predicted my entire career at the age of nineteen. But, it is a great starting point to which I can always refer back. I have listed the most important aspects of what will truly make me happy and if I can follow these simple plans, my life will be successful. On that note, I will consistently alter my plan as my needs change as I grow both personally and professionally.

(James Sadey, Electrical Engineering Student, Fenn College of Engineering, Cleveland State University)

Let us give you a few observations, make sure that your goals are things you can control. James said one of his goals is to maintain a 4.0, which we are sure he will. However, grades are controlled by faculty members, so James would have been better off stating his goal is to do those things that will maximize the likelihood of his earning a 4.0.

Chapter resource 13.3

Additional Career Plan Examples

SAMPLE CAREER PLAN #1

VISION

Personal

Enjoy a lifestyle that brings happiness to me and my family, and makes a positive impact on my local and professional communities. Establish the ability to care for my family and take advantage of many opportunities with no regrets on decisions.

Professional

Focus my career on mechanical engineering as it relates to vehicles. More specifically, apply a variety of mechanical functions to create one great product. In addition, research the conversion of energy in vehicles and apply new ideas for energy conservation.

Enter the automotive industry and join a team of designers with a goal of finding a more environmentally friendly car. Then, secure a management position in research and make decisions to benefit the economy and the country.

Play a major role in the production of a new product and make enough money to support myself and my family and to give back to the community.

Strategy

Goals

Long term

- Develop a lifestyle to bring success and happiness for all members of my family.
- Be financially independent and debt free.
- Be a leader in my church, community, and school system and use my mechanical engineering skills to enhance the secondary education system.

- Provide for the education of my children and help them start their careers.
- Have a management position with a positive impact on my career and profession.
- Prepare for an enjoyable and successful retirement.

Midterm

- Start a family and become active in my church and local community.
- Expand professional networks by membership in engineering societies.
- Plan time with family to develop quality time and enhance the enjoyment of life.
- Establish a budget and savings plan as a financial foundation for my vision.
- Finish school with a bachelor's degree in mechanical engineering.
- Complete graduate school with a master's degree in mechanical engineering.
- Gain industry experience while in school.
- Start working career in automotive industry.
- Become a licensed professional engineer.

Short term

- Learn to balance time for school, work, and leisure.
- Build a personal network of friends and classmates.
- Develop effective study habits to obtain maximum benefits from classes.
- Evaluate the mechanical engineering profession and decide what elective classes will provide the best method to achieve my professional vision.
- Establish an initial GPA of at least 3.80 and develop a process to maintain a high GPA for entry to graduate school.
- Research different engineering colleges and determine if the present curricula is appropriate to prepare me for my vision.

Skill Needs

Technical

- Gain proficiency in calculus and differential equations.
- Increase knowledge and application of engineering software.
- Enhance "hands-on" experience in car maintenance.
- Learn how to apply the principles of thermodynamics and heat transfer.

Nontechnical

- Develop library and other information research skills.
- Improve grammar and writing skills.
- Develop oral presentation skills.
- Learn how to work with other people.

Action Plan

Short term

- Develop a time-management system and learn how to balance time between school, professional societies, and part-time work.
- Join the student chapter of SAE and work on the Baja Car Project.

- Analyze each class to determine how it helps to prepare for my desired work focus and work to obtain a maximum benefit from each class.
- Join a study group and to learn how to work with other students.
- Meet with advisor and review class options to plan my undergraduate class schedule.
- Allocate 30 hours/week for textbook review, homework problems, and test preparation.
- Attend all classes and spend extra time in preparation for exams to receive high scores and grades to support a high GPA.
- Use the library and Internet to research various university engineering programs for direction in completing undergraduate and graduate degrees.
- Establish a tutor for calculus and increase time to practice solving problems when involved in the Differential Equations class.
- Use the University Writing Center for suggestions to improve writing skills.
- Be an officer in SAE and practice leadership, interpersonal, and communication skills.
- Apply for a co-op position in the automotive or related industry.

Midterm

- Get married and start a family.
- Participate in community activities when my children start school by working with science teachers to encourage students to consider engineering careers.
- Be an active member of my church and develop a religious foundation for my family.
- Be an active member of SAE and ASME and participate in activities relating to the automotive industry.
- Work with my spouse and establish quality time for all family members.
- Plan a family budget and use of funds that will provide the desired lifestyle and build a savings foundation for future use.
- Use vacation time to plan and enjoy travel and other activities that will educate and provide recreation for all family members.
- Establish a physical fitness routine to maintain good personal health.
- Continue to develop and use efficient time management and study skills to maintain a high GPA and to gain the most from each class.
- Expand personal and professional networks by active membership in my community and my professional societies.
- Complete FE exam during senior year in preparation for registration.
- Select university for graduate school.
- Complete undergraduate and graduate school.
- Secure my first position in the automotive industry.
- Complete PE exam to become registered as a professional engineer.

Long term

- Work with my family to plan activities and quality time together.
- Support my spouse's career and other activities to build a solid marriage.
- Work with my spouse to review and make successful investments and take advantage of other financial tools such as the 401k program to build security for retirement.

- Consider being a school board member or using volunteer opportunities to improve the secondary school system in my community for my children and other students.
- Develop a plan to help my children review career opportunities and select their life work.
- Provide partial or full financial support of higher education opportunities for my children.
- Develop a lifelong learning program that includes preparation for management and other career positions.
- Seek and secure career related positions of authority and leadership to contribute to long-term goals and professional vision.
- Continue a physical fitness routine to support quality of life and extend retirement years enjoyment.

SAMPLE CAREER PLAN #2

VISION

Personal

The primary focus of my personal vision is on my family. I would like to provide financial support so my wife can be a full-time mother. In addition, it is important that I have the opportunity to be at home and spend quality time with our children.

I would like to have a career that supports a reasonable lifestyle but money is not the most important thing for me. It is much more important to enjoy a happy and successful family and not be concerned about having a lot of expensive gadgets.

Professional

Initially, I want to teach in an undergraduate civil engineering program at a well-known university. I am willing to work hard and learn about the engineering education process because I would like to eventually be a dean of engineering.

My desired working environment is one that includes the association with many groups of people. I enjoy using my communicating and socializing skills to work in teams and enjoy the benefits of different ideas and viewpoints of others.

I would be happy if my career accomplishments include working on projects to improve or save lives. This would give me satisfaction by making a positive impact on my world.

It is important to make enough money to care for my family, help my community, and contribute to the well-being of other people in this world in a similar way that others have helped me.

Strategy

Goals

Long term

- Enjoy life and activities associated with my family.
- Work with my children to help them do well in secondary school and to prepare for their vocations.

- Spend quality time with my family.
- Expand my professional network and practice leadership skills in preparation for seeking a position as dean of engineering.
- Learn how to use financial resources efficiently and to be satisfied with a reasonable lifestyle.
- Find opportunities to be involved in my community and to participate in various activities that provide positive impacts on the community.
- Be a dean of engineering and encourage students to do well in their preparation for their career.
- Be prepared for retirement and help my children enjoy successful lives.

Midterm

- Find a wife, get married, and start our family.
- Develop a balance between school and family activities.
- Establish a good time management system to reduce stress and improve my ability to have extra time for my family.
- Complete my bachelor's degree in engineering and prepare for graduate school.
- Determine the main focus of my career and how I can incorporate helping other people in research associated with teaching in a university.
- Complete a master's degree in engineering.
- Learn how to work efficiently with other people and become a leader.
- Obtain a doctorial degree and become a teacher.

Short term

- Learn how to be a successful student and make the most of class opportunities.
- Obtain good grades in all classes and keep a GPA of 3.75 or better.
- Learn more about academia and what is involved in teaching and becoming a dean of engineering.
- Develop a personal network and learn how to work better with other students.
- Research various engineering fields and make the final selection of the type of engineering to study and receive a bachelor's degree.

Skill Needs

Technical

- Learn how to use computers and software.
- Increase knowledge in math and science theory.
- Enhance my skills in solving problems.

Nontechnical

- Develop leadership skills.
- Improve communication skills.
- Learn how to influence other people.

Action Plan

Short term

- Attend all classes, read the textbook before class and ask questions in class to obtain clarification or additional information.
- Be optimistic and spend at least 2 hours to prepare for each exam.
- Spend time in the library and with other resources to learn more about academia and what is involved in teaching and administration.
- Talk to the dean of engineering and instructors and find out what they do on a daily basis and what they did to prepare for their present positions.
- Form a study group and work with other students a minimum of 5 hours each week.
- Join a student professional society and volunteer for a few projects.

Midterm

- Expand my personal network and share many activities with friends so I can find the love of my life, get married, and start a family.
- Improve my time management system to be more efficient and have more time for my family as well as the balance of my education.
- Work hard, plan daily activities, and maintain a high GPA so I can graduate and enter the graduate school of my choice.
- Select the area of interest and complete a master's degree in engineering
- Continue active membership in my engineering professional organization and expand my professional network and leadership skills.
- Complete my doctorial degree and prepare for academia.
- Learn how to handle failure and build success from experience.

Long term

- Work with my wife and plan activities that will be enjoyed by the entire family including trips and other vacations.
- Discuss career opportunities with my children and be involved in their secondary educational activities to encourage them to be prepared to make good career choices.
- Take advantage of opportunities for leadership positions in my professional society and expand my professional networks.
- Develop a family budget and prepare for future expenses such as college for my children and retirement.
- Review research opportunities and select those that will make a positive impact on society to fulfill my professional vision.
- Prepare and seek opportunities to be a dean of engineering.
- If successful in securing a position as dean of engineering, use my vision to encourage faculty to provide a quality educational environment and students to prepare for and compete a successful career experience.

SAMPLE CAREER PLAN #3

VISION

Personal

The most important part of my personal vision is to provide financial and personal support to my family. My desired lifestyle is to focus first on my family, then my work, and finally, on my community. So, I want to plan and control my career to enjoy a good balance between work and a personal and family life.

I would like to be a wife and mother who is respected and loved. Enjoying a close family is very important and I want to have a home environment that will make that possible. This includes spending time every day with my family and using vacation to introduce my children to new ideas and experiences.

Because I am also interested in my community and society in general, I would like to make positive contributions through my profession. This includes becoming a professional engineer and being ethical in all that I do.

It is important that I maintain good health so I can be successful at work and be a good support for my family needs. I would like to retire in good health so this part of life will also be enjoyable.

Professional

My primary area of interest is chemical engineering and environmental related equipment for buildings. I would like to start my working career with a mid to large size consulting firm where I would get the chance to work on a variety of different projects such as healthcare, educational, industrial facilities, and the like.

My desired work environment is to spend at least 40% of the time in the field and participate in the application of my design projects. I would not be happy in an office all the time.

In addition to obtaining a broad range of work experience, I want to build a large professional network with others in the industry. This will be an important resource for me to be successful the second part of my career.

I want to work for an existing company for about 10–15 years in order to develop the necessary skills and experience to start and run my own business. My long-range plans include establishing my own engineering consulting firm where I would specialize in the design of environmentally friendly systems for major commercial buildings.

Goals

Long term

- Initiate the process of starting my own engineering consulting firm.
- Specialize in environmental related systems for commercial facilities.
- Assume leadership roles in local AIChE and other industry-related organizations.
- Start with light commercial system designs and progress to larger-scale projects.
- Provide financial support for my children's education and my future retirement.
- Balance time between family and business and enjoy vacations with my family.

Midterm

- Be employed by an engineering firm, working on my 4 years of work experience needed in order to take the PE exam.
- Complete payment of all student loans.
- Be active in AIChE and expand my personal and professional networks.
- Develop a business plan before my 35th birthday and start preparations for owning a consulting company.
- Establish a solid personal financial plan to support my family needs and to prepare for company ownership.
- Continue my personal exercise program to maintain good health.

Short term

- Have an outstanding spring semester in order to improve my GPA and postgraduation marketability.
- Complete a successful Senior Design Project and graduate before my 25th birthday.
- Increase my activities in the student branch of ASME, build a personal network of peers, and practice leadership and interpersonal skills.
- Pass the Fundamentals of Engineering exam before graduation.
- Take the PE exam and become a professional engineer before my 30th birthday.
- Develop a personal exercise program.

Skill Needs

Technical

- Increase my ability to process complicated information more efficiently.
- Stay up to date with the most current technologies necessary to be competitive.

Nontechnical

- Learn to balance my work and personal life in order to maintain an adequate level efficiency.
- Learn to utilize the available resources that I have taken for granted in the past (i.e., Careers Services, AIChE contacts, etc.).
- Find a business associate with marketing skills and contacts to help establish and maintain a successful consulting company.

Action Plan

Short term

- Use time wisely and efficiently by expanding my time-management system and reviewing results on a daily basis.
- Learn to say "NO" to extra activities that would interfere with my primary objectives.
- Organize a senior design team and select a project that can be successfully completed within the time allocated.
- Attend monthly meeting of AIChE and volunteer for a limited number of activities in which I can expand my network and practice communication and leadership skills.
- Prepare for the FE exam by allocating 3 hours/week to study exam topics.

- Document projects continue a learning program that will prepare me to successfully pass the PE exam.
- Establish an exercise program that includes running at least three times/week at Bally's or on the streets in my neighborhood.

Midterm

- Increase my marketability by utilizing the contacts and resources made available (Career Services, Engineering Networking Events, etc.).
- Work hard on my initial job position after graduation to establish a solid foundation of expertise.
- Expand my knowledge of the overall operations of a consulting firm by talking to several departments in the company.
- Control finances responsibly to pay off my student debt and increase my net assets.
- Join AIChE and volunteer for leadership opportunities to further expand my professional networks and project management skills.
- Use SCORE seminars and develop a financial plan for my consulting company.

Long term

- Research opportunities for location and type of consulting services to make successful decisions on starting my company.
- Develop adequate knowledge about risks and liabilities involved with owning a business by contacts in AIChE and by conversations with other consulting engineers.
- Initiate the development of a solid financial backing and support from my family and from financial institutions by completing and discussing a business plan.
- Establish a part-time business and explore opportunities for expansion of my consulting business while working for my employer.
- Find at least three to five potential clients who are willing to sign contracts with my company and complete designs of their projects.
- Review results of early operations for my company and decide if I should leave my employer and operate my own business on a full-time basis.
- Use time management to develop a balance between my business and family.
- Provide at least 75% of the expenses for a 4-year undergraduate education for each of my children and help them decide on schools and career paths.

SAMPLE CAREER PLAN #4

VISION

Personal

The most important part of my personal vision is first to be successful in my academic career. I want to make my family proud, especially my dad who is sacrificing so much for me to be in college.

After obtaining a bachelor's degree in electrical engineering, the next step is to find a job in America that will provide experience in the communications field. This work experience will be important in developing skills and in building a good resume so I can return home to Kenya and continue my career. In addition to desiring a successful career, it is important that

I live close to my parents and start my own family. As all this unfolds, I want to develop as a man and make the right decisions in my life.

I will dedicate all my knowledge acquired in college to help the company I find myself working for while at the same time investing in my father's real estate business in Kenya.

A healthy physical, social, and financial state is also another priority. I want to be a healthy man for as long as I live. I know this is hard, but I want to eat right and exercise enough. It would be nice to be wealthy, but being financially stable is good enough for me. Socially, I'd like to make good new friends and keep in touch with current friends.

Professional

My major career focus is communication systems. As an electrical engineer, I'd like to work for a company involved with communications, such as Ericsson, Samsung, and NASA. Working for this type of company would provide important hands-on experience. Because there is a lack of efficient mobile communication systems and Internet availability in my homeland, I would like to start my own company to compete with the few that monopolize this industry in Kenya.

Initially, the most desirable work environment would be a large corporation. Then, after gaining experience, I will move to Kenya and work to establish my own company.

If successful in establishing my own company, it will be important for it to grow so I can provide excellent services and make a positive contribution to society by developing and advancing communication technology.

Strategy

Goals

Long term

- Have a successful business of my own in Kenya.
- Enjoy a simple life without worry or stress by building financial security.
- Help my children obtain a college education.
- Financially support my children and help them start their own careers.
- Take an active role in IEEE and my community and help my neighbors improve their lifestyle through the application of technology.
- Retire and enjoy spending time with my grandchildren.

Midterm

- Complete classes and receive a bachelor's degree in electrical engineering.
- Obtain a position in an American-based company in communications.
- Establish my family and spend quality time with them.
- Minimize stress and enjoy a balance of work and family.
- Take care of my parents in their older years.
- Develop a healthy lifestyle to minimize health problems.
- Move to Kenya and start my own communications business.
- Establish an IEEE Section in my community in Kenya.
- Provide employment opportunities for people in my community.
- Be a silent partner in my father's real estate business.

Short term

- Take personal responsibility for my activities and learn to manage time.
- Complete all tasks in my time-management plan to be more successful.
- Expand personal networks and search for a future wife.
- Join IEEE and be active in projects to learn how to work with others.
- Start a physical fitness program and exercise on a routine basis.
- Increase my ability to obtain more technical knowledge from each class.
- Find part-time work and add work experience to prepare a good resume.

Skill Needs

Technical

- Learn how to apply mathematics to each engineering class.
- Improve my ability to solve homework and test problems.
- Learn the principles associated with wireless communication systems.

Nontechnical

- Learn how to be more effective in preparing for classes.
- Learn how to manage time and to minimize stress of time schedules.
- Be physically fit.
- Strengthen my religious faith.
- Learn how to be patient and to work with others who have different ideas.

Action Plan

Short term

- Develop a time-management system and revise it at the start of each day.
- Review textbook material and attend all classes.
- Write questions when reading textbooks and ask them in class to better understand the technical material.
- Join a study group and learn how to work with others in solving problems.
- Walk four miles on Monday, Wednesday, and Saturday each week.
- Apply for a part-time job in the University Student Center.
- Attend church every Sunday.
- Join IEEE this semester.

Midterm

- Schedule conferences with my advisor and select elective classes to prepare to work in the communications field.
- Write a brief summary of what I can obtain from each class and improve my technical skills by working homework problems with my study group.
- Learn how to work in a laboratory team and how to use equipment to conduct successful experiments.
- Volunteer for leadership opportunities in IEEE and learn how to lead committees and manage projects.

- Start my working career in a company in America.
- Increase my activities in church and expand my personal network so I can search for a future wife.
- Work with a personal trainer and expand my physical fitness program.
- Send money to my father so he can invest this in his real estate business.
- Focus on obtaining experience in the application of communication technology in preparation for relocating to Kenya.
- Get married and start a family.
- Start my own wireless communication company in Kenya.
- Work with my parents and help them plan for retirement years.

Long term

- Use good business principles and operate my company so that it is successful and provides a good living for my employees.
- Develop a budget and establish a saving plan for expenses for my children's college.
- Be a leader by volunteering in community activities and by helping the schools improve their math and science classes.
- Establish a management team that can continue successful operations of my company and provide retirement benefits to me and others in retirement.
- Help my children start their careers and plan time with my grandchildren.
- Encourage young people to improve their overall lifestyle for our community by offering to tutor them in math and science subjects in high school.

CHAPTER

14 Networking and Professional Development Skills

In Search of Diamonds

Career success is often based more on who you know than on what you know. Developing effective networking skills now, as a student, is one of the most important nontechnical tools you can possess. Then, by continuing to build your network of contacts as a practicing engineer, you can use your contacts to obtain important information that will greatly increase your level of success.

—Joseph Burns, PMP, MS Aeronautical Science,
BS Astronomy & Astrophysics

NETWORKING AND PROFESSIONAL DEVELOPMENT SCENARIO

"As we begin the development of our professional career we must recognize the need for interaction with professionals in our area of interest, and we must understand the outcomes that can and will result from this interaction. We must look for opportunities to meet with these individuals and to learn from those who have many years of technical experience.

We must also recognize the need for nontechnical skills and identify the specific skills that will be required if we are to be successful in our technical career. While technical knowledge and ability are requirements for success, we must also understand that there are several nontechnical skills that will assist us in our professional career."

(Joe Lillie, Electrical Engineer, former IEEE Vice President, and IEEE Foundation Director and Treasurer)

As a professional engineer with many years of experience, Joe Lillie knows the value of networking. In this chapter, we discuss how you can develop your professional networking skills to achieve greater career success. Then, we learn more about Joe's networking experiences when we visit with him at the end of this chapter.

LEARNING OBJECTIVES

By using the information and exercises in this chapter you will be able to

- Understand why personal networks are important.
- Make positive first impressions.
- Develop the valuable skill of listening.
- Know how to build networks.
- Establish networks in classes and study groups.
- Use student professional organizations to expand networks.
- Know how to use professional organizations to enhance your career.

14.1 Introduction

There is a myth that engineers tend to be antisocial and work in isolation. Nothing could be farther from the truth. Most engineers establish personal contacts and exchange ideas with technical and nontechnical people. This is called networking, and it is one of the best methods to obtain information and enjoy successful careers.

You will never be an island.

Engineers do not operate in a vacuum or as an island. As indicated in Figure 14.1, we are in the "people" business. Our ability to *share* information is just as important as the information. In addition, we need many sources of information beyond our own knowledge. Developing contacts and building networking environments open large areas of resources for our benefit. So, networking is valuable, like diamonds, and is a skill that will help you learn about new technology.

The center word in networking is "work," and it is true that you need to work to develop effective skills to interact with others. This starts with the desire to be involved with many people in your personal and professional communities. If you have a positive attitude about networking, you will be successful in creating effective networks.

Some networking opportunities are initiated when others contact you. This helps to build your network, but the best contacts are often the ones you initiate because you select whom you want to meet. Even if you are a little uncomfortable meeting new people, you should consider the benefits of developing effective networking skills.

Figure 14.1
Networking.

"Young people have the innate ability to learn, observe, explore, and acquire wisdom. They hope to achieve success in their meaningful careers. One can study specific areas and become a subject matter expert; but to achieve one's full potential, one must develop personal attributes that enhance an individual's professional development and networking skills. Faced with the challenges of the 21st century, students design products that can have tangible impact in the areas of health care, energy, transportation, and other sectors to benefit society. The ability to communicate innovative ideas, partner and collaborate effectively is critical in maximizing job performance and improving career prospects moving forward."

(Carole C. Carey, Electrical Engineer, Chapter Chair IEEE Engineering in Medicine and Biology Society, Baltimore Section)

14.2 Successful First Impressions

Many years ago, Will Rogers said that we never get a *second* chance to make a *first* impression. This statement is just as true today and should be the foundation for our attitude when meeting new people. A good way to make a positive first impression is to use friendly body language as shown in Figure 14.2.

Little things make a big difference when you start a conversation. This includes standing when someone greets you and giving a firm but not too strong handshake. Positive body language can also be demonstrated when you smile and establish eye contact.

Most first impressions are made in 15 seconds.

Figure 14.2
Positive first impressions.

Typically, it takes about 15 seconds to make a first impression. If it is negative, it may take a very long time to undo the damage. But, if it is positive, you will establish a foundation for a successful discussion. As many as twenty opinions about a person can be established in the first minute or so of a conversation.

A positive first impression, when talking to prospective employers, is often a major determining factor in receiving a job offer. Some recruiters indicate the opportunity for a second interview or job offer is based on social skills displayed during the first meeting. You need more than technical knowledge and good grades to be competitive during job searches, so learn now to make a positive impression and this will be a valuable asset.

14.2.1 Initial Conversation Tips

Start a conversation with friendly comments.

The first few words of a conversation are very important. One of the best methods of starting a conversation is to smile, say hello, and introduce yourself. Look the other person in the eyes and state your name slowly and clearly.

If possible, make a statement about your name to help the other person remember it easily. For example, "Hello, I'm Jim Watson—just like the Watson who received the first telephone message from Alexander Graham Bell."

Effective communication skills can be one of your most important professional tools. The art of conversation includes knowing *what* to say and *when* to say it. It also includes knowing *when to be silent*. By developing the art of conversation, you will put others at ease and be more successful in your first impressions.

YOU CAN USE QUESTIONS TO

- Direct the topic of discussion.
- Compliment the other person by indicating you are interested in his or her ideas.
- Transfer the focus from you to them.
- Target areas for comments or further questions.
- Build self-confidence as you listen before speaking.

Figure 14.3
Value of asking questions.

14.2.2 The Power of Questions

When starting a conversation, it is good to ask questions after identifying yourself in the initial introduction. When you ask appropriate questions, you toss the ball of conversation to the other person. Questions can be a valuable tool to accomplish the items listed in Figure 14.3.

Make initial questions open ended to encourage others to expand their comments. An open-ended question cannot be answered with a yes or no. Open-ended questions target the topic, but do not limit the response. The end result is that you receive more information and also have a better idea of where to go next in the conversation.

Open-ended questions are a powerful tool in networking.

Open-ended questions provide information to guide your next statements or questions. Listen to responses and use questions or statements to continue the conversation. You can accomplish this by nodding in agreement, maintaining good eye contact, listening for key words, and rephrasing statements to indicate you understand. We have two ears and only one mouth, so learn how to be an active listener and respond to what others say.

Be genuinely interested in people and their ideas. Because you are looking for new information when networking, respect others and be ready to consider different viewpoints and diverse backgrounds. An open mind will be more receptive to new ideas even when you may not initially agree with some concepts. Be open to new ideas at first, and then decide if or how the information will apply to you.

Be concise when you answer questions so you do not overwhelm others. When someone asks what time it is, don't tell them how to make the watch. On the other hand, you should not be so brief in your comments that you sound curt or that you are uninterested.

14.2.3 Changing Discussion Topics

There are several methods for you to take control of conversations without negative impacts. It is important to be open and friendly, so use tact to change the subject.

You can also use bridge statements to move the conversation to a new topic. For example, start the transition with statements such as "I heard you mention . . ." or "You seem to be familiar with . . ." and then introduce the new topic you mentioned in your transition statement.

Bridge statements show other people you have been listening to them before you want to change the subject. This is a polite and effective way to encourage others to discuss new topics that you initiate.

14.2.4 Body Language

Negative body language creates barriers between you and other people. This can interfere with your ability to enjoy a productive conversation. So, think about how you appear to others, and develop body language to make a positive impact on people during conversations.

Use a positive attitude to build a foundation of trust and to encourage others to be comfortable when joining in conversations. This can be demonstrated with a smile and other positive body languages techniques.

Posture and appearance also impact initial impressions. Good posture demonstrates a sense of self-confidence. Your personal appearance is an indication of self-esteem and indicates your desire to make a good impression.

14.2.5 Ending the Conversation

End the conversation on a positive note by smiling and expressing thanks to the other person for talking to you. Except for close friends, a parting handshake is the *only* appropriate form of touch in business. A pat, nudge, or touch on the arm may be perceived as being too friendly.

14.3 The Art of Listening

> *"To listen well, is as powerful a means of influence as to talk well, and is as essential to all true conversation."*
>
> (Chinese Proverb)

Because listening is such an important part of networking, let's focus on this topic from a larger perspective than just one-on-one discussions. Listening is an active process of receiving and processing information in many different situations. If you think about what you hear, you can understand what is said and retain the information much longer.

> Different levels of listening are casual, conversational, and comprehensive.

14.3.1 Casual Listening

Casual listening is primarily for enjoyment. This includes music, radio, television, and other leisure activities. Opportunities for casual listening have been greatly expanded by portable devices.

14.3.2 Conversational Listening

Conversational listening involves informal discussions with other people. New ideas can be learned in informal conversations, and the amount of information obtained depends on your active listening skills.

14.3.3 Comprehensive Listening

Our main focus in this chapter is on comprehensive listening because this involves understanding what others say and giving feedback. This includes using positive body language, paraphrasing what you hear, and asking questions for clarification.

Comprehensive listening requires some work on your part, but you have many opportunities during your educational experience to develop listening skills in discussions with other students, instructors, and friends.

You need to use the highest level of listening skills to understand information in lectures and other group meetings. In contrast to casual and conversational listening, comprehensive listening involves more concentration on the details of what you hear.

As a student, you can save time and complete work more accurately when you listen to details. When receiving class assignments, listen carefully, process what you hear, and ask appropriate questions. This will help you understand assignments, and you will learn skills to apply in future career projects.

Comprehensive listening often includes taking notes for additional reference. It is almost impossible to write everything that is being discussed and still be thinking about the material. So, select the most important points to record, and balance your time between writing and listening.

Engineers frequently attend meetings in which speakers discuss various topics, as shown in Figure 14.4. This is an excellent way to learn new ideas and to keep up-to-date with changes in technology. The value obtained from these meetings depends on how well you listen and comprehend information.

Comprehensive listening tools

- Attitude
- Distractions
- Emotion
- Enthusiasm
- Difficult Topics
- Message
- Main Ideas
- Notes
- Hidden Meanings
- Conclusions

Figure 14.4
Listening to formal presentations.

Comprehensive listening maximizes the value received when you attend meetings.

14.3.4 Comprehension Listening Enhancement

Maintain a Positive Attitude—Even if the subject is not interesting, accept the challenge to obtain the most from the discussion. Focus on the speaker's message and try to get something of value from each speaker.

Control Distractions—It may be difficult to eliminate distractions, but you can minimize their impact. For example, you might want to sit near the front of a large audience. Distractions will have less impact if you focus on the speaker and the message, and not the environment.

Minimize Emotion—If some topics stir your emotions, try to minimize your feelings during listening experiences. Strong emotion can block understanding and may lead to incorrect conclusions.

Listen with Enthusiasm—Listening is not an easy task. It requires interest and attention. Observe nonverbal as well as verbal messages to better understand the true meaning of the message.

Practice with Difficult Topics—Guest speakers for professional organizations often provide challenging topic discussions. These offer opportunities to practice listening skills. Focus on the main ideas and make the most of this learning experience.

Focus on the Message and Not the Speaker—Some speakers have little experience or are less than professional in their delivery. In these situations, information in the presentation can be useful if you focus on the message and not speaking skills.

Focus on Main Ideas—Presentations often contain a large amount of information that supports main ideas. Focus on main ideas to prevent distraction and use supporting information to better understand the main topics.

Take Appropriate Notes—Writing notes is a good way to gain more from the presentation if this does not distract from listening. Typically, notes should be limited to main ideas.

Look for Hidden Meanings—Speakers often have hidden agendas or strong biases that may not be demonstrated by their words. Observe the speaker's body language and vocal expressions so you understand more than just spoken words. This requires more work on your part, but it is a good way to get the most from the presentation.

Don't Evaluate Too Early—Wait until all of the information has been provided before making conclusions. Early conclusions interfere with listening, and may result in missing important ideas later in the presentation. Without considering the total information, you may form incorrect ideas.

14.4 Opportunities to Practice Networking Skills

The value of networking is in the information you receive and process. You can understand more when you "read between the lines" to determine why something is said. Knowing the reasons comments are made may be just as important as the comments.

Listening is also important in networking. Networking is most successful when you maintain eye contact and use positive body language. This demonstrates that you are listening and understanding what is being said. The more you listen the more you will be viewed as a good communicator.

Focus on others when networking.

To build long-term networks, be sincere. Some comments can be used to give compliments to others, but these should always be based on fact and be a sincere expression of your appreciation for the knowledge or action of others. Insincere flattery will usually be received in a negative manner.

14.4.1 Classes, Laboratories, and Study Groups

As a student, you have many opportunities to practice networking in low-risk environments. Classrooms offer opportunities to network with instructors. You can build effective relationships with instructors by talking to them before and after class. And, you can expand your contacts with students. This should be easy because of your common interests.

Laboratory teams offer excellent opportunities for you to practice networking. When you are part of a team, you will soon learn that other members often have different ideas about how to conduct an experiment. When you listen and discuss these ideas, you can build networking skills as well as technical skills.

Figure 14.5
Networking in study groups.

When you participate in study groups, the synergy of brainstorming provides an opportunity to solve problems efficiently. Perhaps even more importantly, study groups offer opportunities to practice networking and develop confidence in working with others, as shown in Figure 14.5.

Future work assignments will often include involvement in committees, project teams, and other groups. So, by learning now to work effectively in a study group you will be preparing for similar group activities in your working career.

14.4.2 Professional Organizations

But networking is not limited to classrooms, laboratories, or study groups. Professional organizations, such as ACM, AIChE, ASCE, ASHRAE, ASME, CSA, IEEE, NSBE, SHPE, SAE, SWE, WIE, and many others, provide technical information and offer many opportunities to become acquainted with a variety of students and professional engineers.

> *"I am always looking for networking opportunities. In my experience I have found that networking is the most effective way to develop relationships that build business and assist in the advancement of my career.*
>
> *Joining a student organization is one of the best ways to start networking. Student organizations provide some of the best opportunities to network and develop relationships with professional mentors that will share their experiences and help you develop a strategy to successfully manage your career."*
>
> (Maria Marez Baker, BEE, Technical Advisor, Electrical Arts, Frisina, LLC)

The best way to develop networks in professional organizations is to be an active member. You can start now by attending meetings and volunteering for opportunities to practice networking by being involved in existing projects. Then, you can suggest new projects or activities and lead or participate in project implementation.

Activities in professional organizations help build networks with upper classmates while they are in school. Maintaining contacts with peers after they graduate can lead to important connections within industry and regional groups.

"Becoming actively involved in professional societies like IEEE—not just showing up, but planning activities and leading organizations—played a huge role in shaping my college education beyond what I learned in the classroom. Academic performance shows commitment to the technical skills required of engineers, but networking and society membership shows commitment to the engineering profession itself. When it comes to getting a job, I've learned that grades may get you noticed, but networking gets you remembered."

(Greg Back, Computer Engineer and Information Security Engineer at MITRE)

The value of networking does not stop upon graduation. By continuing active membership in professional organizations after graduation, you can expand your contacts to regional, national, and international levels. This will result in an ever-expanding network as a tool to increase your opportunities for future career growth.

14.4.3 Electronic Networks

Another excellent opportunity to practice networking is to use one or more electronic networks. Establish a list of friends, classmates, instructors, practicing engineers, and other professional contacts during your student experiences. Then, this can be an effective method to continue communications with important contacts after graduation.

Electronic networks offer an easy way to exchange messages on many different topics with your family, friends, and professional peers. When sending electronic messages, remember these can be viewed by many individuals.

For example, recruiters can read your messages on electronic networks when considering your qualifications as a future employee. So, think about what you say on your networks and consider how this may impact future employment opportunities.

A CONVERSATION WITH JOE LILLIE

Joe, you are a good example of an engineer who uses networks. How did you learn to do this?

Answer:

"Nontechnical skills are developed by observation and practice. During my professional career I was able to interact with many technical professionals and this interaction provided me the opportunity to observe the skills that these individuals displayed.

I was also able to initiate one-on-one discussions with many of these professionals and I used these discussions to better understand how the various skills were developed. I was able to use these discussions to identify the specific skills that I could develop and use to improve my technical performance.

Over time it became clear to me that I could use my participation in nonwork events such as social clubs, professional organizations, and parent/teacher groups to practice my various skills and gain the confidence to use these skills in the work environment.

The other important lesson learned was that I could ask close friends and family members for feedback on my skill usage and then use the constructive feedback to improve my skills application."

What nontechnical skills do you use to develop contacts and to network?

Answer:

"Many of the following skills have been used in building my networks:

- Oral Communications—the ability to properly communicate with others in a way that they clearly understand the information that you are attempting to convey.
- Written Communications—the ability to document technical issues such that the identified concepts and processes are clearly understood by the reader.
- Presentation Skills—the ability to properly convey a specific message in front of large groups and the ability to properly manage the audience.
- Negotiations—the ability to fully understand the issues related to a specific topic, recognizing the implications of certain decisions and seeking a win-win outcome.
- Teamwork—the ability to work with others to complete group tasks, seeking input from all team members and respecting the diversity of all team members.
- Good Manners—conducting oneself in a professional manner and showing respect for others.

Applying nontechnical skills in a nonwork environment is the best way to improve your application of these skills. This can be accomplished by delivering presentations to groups outside of the work environment. These presentations can be somewhat technical but presented in a nontechnical manner.

Using this approach you will actually be utilizing several nontechnical skills since you will have to organize your thoughts, document these thoughts in your presentation material, and then actually make the presentation. During the presentation you will be using oral communications to convey the information.

You will also have to manage the audience during the presentation and as you field questions from the participants at the conclusion of the presentation. You can also seek feedback by preparing a short feedback form that you should ask each audience member to submit onsite after the presentation is completed.

Providing the participants with an email address will allow for additional feedback on the presentation content, presentation style, and help you build your professional network.

You will have to do a little work to get on the agenda for the various meetings but the dividends derived from this process will greatly exceed your efforts."

What advice would you share with students?

Answer:

"I would make three recommendations to the students to achieve greater career success:

1. Join and become active in your professional society. Active involvements in organizations such as the Institute of Electrical and Electronics Engineers (IEEE) and the American Society of Mechanical Engineers (ASME) will assist you in developing technical and nontechnical skills while providing you with access to professionals in your area of interest.
2. Develop your nontechnical skills. Identify the specific skills required for your current and desired work assignments and assess your level of performance of the activities associated with these skills. Seek input from your supervisor(s), close friends, and family members. Ask them to give your their perceptions of your performance related to the various skills. Use this valuable feedback to identify gaps in performance and seek opportunities to improve by attending training sessions, seminars, and conferences.
3. Build your professional network consisting of individuals who can assist you with career issues. Every technical professional that you meet can become part of your network so your network will include many individuals. You should take some time to identify the five individuals from your network who can assist you in your time of need. You will at times encounter critical issues which will require you to consult with others and your top five individuals will be available to provide help to you at your time of need."

(Joe Lillie, Electrical Engineer, former IEEE Vice President, and IEEE Foundation Director and Treasurer)

14.5 Conclusion

In this chapter, we discussed listening, networking, and how to make positive first impressions when meeting people. These nontechnical skills are valuable tools to help you prepare for unexpected opportunities and to reach greater success in your career and personal life.

You can read about engineering career options in books and on the Web. This is a good start, but you will have a much better idea of what is involved in engineering when you discuss career experiences with engineers. One way to accomplish this is to be an active member of organizations that interface with practicing professionals.

Initiate discussions with speakers from industry when they give programs at professional organization meetings. Practice good listening skills to learn about their engineering careers. Follow up your conversations by sending electronic or other messages of thanks for their participation in the meeting. Then, they may be good contacts for you if you are interested in future job opportunities with their company.

When you attend fieldtrips to learn more about what engineers do in their working careers, initiate as many contacts within the companies as possible. Introduce yourself and ask for business cards. Most engineers have business cards even in this electronic age. Follow up with a networking message that relates to how learning about that company was important to you.

Your academic and industry network contacts could be helpful when you select the area for your graduate school thesis. This is a practical approach to identify research areas that could be an important resource for your master's degree.

Your options for networking will continue to expand with new communication technology. Many electronic networks are available to help you develop and expand your networks on a global basis. People in your electronic networks are good sources of information for new technology and career opportunities. But remember, what you post in an electronic network is available to many readers in addition to the intended recipient.

END OF CHAPTER REVIEW QUIZ

Select the most appropriate answer to the following statements.

1. The most important value of networking is
   - **a.** To get a good grade in this class.
   - **b.** To use instructors as references for job interviews.
   - **c.** To have contacts for information and job opportunities.
   - **d.** To gain a reputation of being friendly.
2. Making a positive first impression is most important to
   - **a.** Influence others to accept your ideas.
   - **b.** Initiate the opportunity to expand your personal network with new people.
   - **c.** Make sure people don't laugh at you or walk away.
   - **d.** Build your self-confidence.
3. The greatest value of asking open-ended questions is to
   - **a.** Learn details and new ideas.
   - **b.** Know what to discuss next in the conversation.
   - **c.** Show your interest in topics being discussed by others.
   - **d.** Achieve all of the above.
4. The best way to change the subject when talking to others is to
   - **a.** Wait until they take a breath, and then talk about what you want to discuss.
   - **b.** Yawn to show you are tired of their discussion and then starting talking.
   - **c.** Use bridge statements to show you are listening but want to change the topic.
   - **d.** Talk over the other person until they stop and listen to you.

5. Comprehensive listening is the best way to
   a. Process and understand details of a class lecture or other presentation.
   b. Appear to be listening and interested in class.
   c. Minimize your participating in a discussion.
   d. Keep from falling asleep in boring presentations.

6. You should practice networking in laboratory team situations because
   a. It will impress your team members.
   b. Experience working with team members will help you to be more effective in job team situations.
   c. You will be able to convince the team to always do everything your way.
   d. You will get a much higher grade in the class.

7. The most important value of being in a study group is to
   a. Complete assignments and to learn to work with other people who have different ideas.
   b. Get higher grades.
   c. Show others how to complete assignments and solve homework problems.
   d. Achieve all of the above.

8. You can maximize the value of student professional organizations by
   a. Being an active member and participating in projects.
   b. Suggesting new projects and offering to lead teams to complete them.
   c. Attending meetings and practicing networking with new people.
   d. Doing all of the above.

9. The most important reason for you to continue active membership in professional organizations after graduation is to
   a. Impress your supervisor that you are smart.
   b. Have contacts for new technical information and for new job opportunities.
   c. Expand your network to regional, national, and international levels.
   d. Attend meetings and conferences.

10. It is important to practice networking before graduation because you
   a. Can improve skills to help you achieve greater success in your career.
   b. Will have many opportunities to enhance listening and other networking skills.
   c. Can build self-confidence to participate more effectively in job interviews.
   d. Can do all of the above.

EXERCISES TO DEVELOP AND ENHANCE YOUR SKILL SET

Exercise 14.1 Listening self-evaluation exercise.

Picture yourself in an audience hearing a speaker deliver his or her message. Be realistic and honest and identify your listening habits by scoring statements from 0 to 4 as follows:

0 = never, **1** = seldom, **2** = sometimes, **3** = frequently, **4** = always

_______ **1.** I think about my own ideas during a presentation and miss some of the speaker's comments.

_______ **2.** I try to simplify what I hear by omitting details.

_______ **3.** I let my mind wander to topics not being covered during a presentation.

_______ **4.** I allow my attitude about a topic or speaker to influence my evaluation of the message.

_______ **5.** I focus on a particular detail of what was said instead of the overall meaning that the speaker is trying to convey.

_______ **6.** I hear what I expected the speaker to say instead of what was actually said.

_______ **7.** I listen to what others say, but I don't feel what they are feeling.

_______ **8.** I relax and listen passively and let the speaker do the work.

_______ **9.** I make a quick evaluation early in the presentation before listening to what the speaker really means.

_______ **10.** I listen to the literal ideas and do not look for hidden or underlying meanings.

_______ Total Score

Exercise 14.2 Professional organization assignment.
Complete the following:

1. Make a list of student professional organizations for the college.
2. Research the value of active membership in these organizations.
   a. Select one or more organizations and attend meetings to evaluate their activities and potential opportunities for your involvement.
   b. Initiate discussions with officers, leaders, and members and discuss advantages of active membership in their organization.
   c. Research the organization(s) on the Web or in printed materials.
3. Select the organization you feel offers the greatest opportunities for you to enhance your career, both as a student and in your future working experiences.
4. Attend at least one meeting of the organization selected in step 3 and prepare a written report based on the format and comments as shown in the Professional Organization Meeting Report.
5. Write your report using good sentence and paragraph structures (not bullets).

Hand in a printed copy of your report at the start of class on the date established by your instructor and send an electronic copy of your report to your instructor on the same day.

PROFESSIONAL ORGANIZATION MEETING REPORT

Prepared by (Your Name) ______________________________

I. Organization Name, Purpose, and Objectives

Use a paragraph to identify the name of the organization and discuss the purpose and objectives of the organization.

II. Meeting Leaders/Speakers and Topics Covered

Include one paragraph for *each* major participant or speaker to identify their name, appropriate background information, and major points of their presentation.

III. New Ideas

Corporations invest time and money when sending representatives to conferences and meetings. Therefore, they anticipate attendees will obtain new ideas that can be helpful to the company and individuals involved. In this section, briefly discuss new ideas that you learned by attending the meeting. These can be related to the organization or to the topics that were discussed by speakers. Typically, this would include one paragraph for each major idea.

IV. New Contacts

Meetings typically provide opportunities to meet new people. Networking activities can occur before the meeting, during refreshment breaks, and after the formal meeting has concluded. Because people are important resources, business cards and other contact information should be obtained and filed for future reference. Use this section to report new contacts established at this meeting. List their names and methods of contacting them and discuss the potential value of communicating with them in the future. Ask questions if needed to obtain this information.

V. Organization Membership

Discuss the value of active membership of this organization. In a second paragraph, identify whether you are a member, plan to be a member, or do not plan to be a member, and explain the reasons for your decision.

Note: Typical meeting reports for this exercise are one to three pages in length. Use the topics and subtopics listed in this outline. For a professional appearance, use left margin justification and a font size of 11 or 12 points. Use a single space within paragraphs and a double space between paragraphs. Run spellchecker and make corrections before saving your report.

CHAPTER

15 Co-Op/Internship and Other Practical Experiences and How This Can Enhance Career Success

Co-Op/Internship and Other Practical Experiences

"The Fenn Co-op Program (co-op) is a structured, academic program that integrates classroom studies with paid, productive, real-life work experiences. Co-op experiences also allow students to begin their professional careers while in school by exposing them to the various environments in which engineers operate on a daily basis. Typically completing three or four semesters of Co-op, engineering students have found employment with government agencies, industrial manufacturers, consulting firms, and other private engineering enterprises."

—Beverly B. Kuch, Fenn College of Engineering, Cleveland State University

Where do we begin? As students your goal must be to prepare yourself to be one of the best engineers you can be and not to become a professional student. Although many of you want to "get out" as quickly as possible, you need to make sure your skill development and enhancement is complete, otherwise you will lack those things that are necessary for a successful career.

In the process of acquiring skills, many students will have the opportunity to participate in a co-op or internship program (from here on whenever we mention co-op, the more formal experience, please understand that most of what we say also applies to internship). This participation has many benefits. However, it may not be for everyone, so we will help you to determine if you should participate in such a program. For many students, participation in a co-op program will substantially increase their skill development and enhancement.

LEARNING OBJECTIVES

By using the information and exercises in this chapter you will be able to

- Determine whether you should consider participating in a co-op program.
- Determine how to secure a co-op assignment in schools that have well-run co-op programs.
- Know how to seek out a co-op type assignment while attending a school that has a weak or nonexistent co-op program.
- Determine how to obtain the most out of a co-op experience.
- Understand the benefits of attending a school with a mandatory co-op program.
- Understand the benefits of attending a school with a nonmandatory co-op program.

Not only does the co-op experience represent an additional avenue for skill development and enhancement, it is also a significant way to motivate students and to increase their understanding of the "fun factor" of an engineering career. Where appropriate, we strongly encourage such an experience to compliment your educational experiences. In this chapter, we discuss the advantages of such experiences and how to obtain the most out of them.

15.1 Introduction

Historically, the process of mentoring future practitioners is one of the oldest forms of education, especially for learning a vocation. The word that best represents this form of education is "apprentice." When engineering education evolved, the need for such experiences became evident. Today we call this co-op or internship. Co-op programs are so well developed that we can find them at most engineering schools. Although co-op programs can encompass nonengineering programs as well, the ones in engineering are by far the largest and best developed. Some schools have a mandatory co-op program as part of their graduation requirements.

The benefits of participation in a co-op program are

- additional skill development and enhancement,
- a source for self-motivation,
- a better perspective as to what an engineer is,
- opportunities for employment, and
- a good source for additional income during the process of obtaining an undergraduate or a master's degree.

The only real downside to doing a co-op is that it will take you longer to earn your degree.

15.2 Why Even Consider Participating in a Co-Op Program?

At the heart of a co-op program is working with engineers in a company that uses engineers. Because you are working with practitioners on engineering projects, you will be practicing the skills that engineers need to have. You will see how real engineers do their work. You will spend a great deal of time communicating with working engineers. You will be expected to participate in the engineering related to the project that has been assigned. You will need to write reports related to your experiences.

As mentioned in the introduction, there are many benefits to participating in a co-op program. Let us look at some of the benefits.

> The most important benefit of participating in a co-op program is skill development and enhancement.

Another major benefit is that if you are not really sure you want to be an engineer, this kind of experience may help you to decide whether you want engineering as your career. If this indeed is one of the reasons you have decided to participate in a co-op experience, you need to very carefully document your experiences so that at the end of the co-op you can make an informed decision.

> Some of the most important benefits to participating in a co-op program are
> - Achieving skill development and enhancement
> - Finding out if engineering is right for you
> - Gaining an additional source of income
> - Helping you become more motivated
> - Increasing job opportunities with the employers related to your co-op assignments
> - Improving your credentials to improve your opportunities for a good engineering position after graduation

One of the concerns about going to college is how do you fund it. Participating in a co-op experience can also be a good source of additional income. Many co-op assignments pay better than other types of jobs a college student can have.

If you are already in an engineering program, the rest of this chapter will relate to possible co-op experiences at your school. However, if you are a high school student considering going into engineering and want to participate in a co-op experience, you need to go to the websites related to the schools you are interested in attending. Fortunately many of these sites contain additional information about co-op programs and the benefits of participation in such programs.

15.3 How to Become Part of a Co-Op Program at a School that has a Required Co-Op Experience

Even if you are already thoroughly immersed into your program, additional effort on your part will help enhance the experience. First go to the Web for your school's co-op program and see how much you can learn. Try to identify the companies that are part of the school's co-op program and go to their websites. See what you can learn specifically regarding a company's participation in co-op programs and in general about that company.

The next step is to make sure your career plan (see the chapter on career management skills) reflects your participation in a co-op experience. You need to have a clear understanding of the expected outcomes from that participation. In addition you will need to make sure you document your experience especially as related to expected outcomes, in a manner that enhances your portfolio (see the chapter on information capture and portfolios).

At the end of each co-op period, you need to document your feelings about the company and list the pros and cons about working with that company once you graduate. Many times co-op students will have a job offer from the company with whom they have worked. Please make sure you do this right after completing that period since our memories tend to fade quickly.

Also at the end of that period, you should determine if you want to try for a new company or return to the same company for the next period. Your specific co-op requirements may not allow you to return to the same company, you need to consult your school's requirements to find out what the school allows.

15.4 How to Become Part of a Co-Op Program at a School that has a Strong, but not Required, Co-Op Experience for Its Students

Many of you may be at a school with a high-quality co-op program but with no requirement that you have to participate. You need to consult your career plan and use it to help you decide if you should participate in the program. Once you have made a decision, then you will either continue toward graduation without participating in a co-op experience or you will participate. If you have decided to participate, then the best thing you can do is to learn all you can about your school's particular program. Interview students who are participating in the program, speak with your school's faculty members about participating in the program, obtain their views in support of both participating in the program and not participating in the program, and learn as much as you can about the companies that participate in the program.

You then need to review your career plan and make changes where appropriate (see the chapter on career management). You need to pay particular attention to your expected outcomes from the experience. In addition you will need to make sure you document your experience especially as related to expected outcomes, in a manner that enhances your portfolio (see the chapter on information capture and portfolios).

At the end of each co-op period, you need to document your feelings about the company and list the pros and cons about working with that company once you graduate. Often co-op students will have a job offer from the company with whom they have worked. Please make sure you record your feelings right after completing your co-op period since our memories tend to fade with time.

Also at the end of that period, determine whether you want to try for a new company or return to the same company for the next period. Your specific co-op requirements, however, may not allow you to return to the same company. You need to consult your school's requirements to find out what the school allows.

15.5 How to Earn a Co-Op Assignment while Attending a School that has a Weak or Nonexistent Co-Op Program

This will be a distinct challenge, requiring you to do much of the work on building the experience and making sure it happens. There are a lot of resources you can pursue. First, we recommend you learn as much as you can about co-op. You should start at the websites of schools that have required co-op experiences and schools that have strong but not required co-op programs. Their websites will supply you with helpful information. Then you need to take this information and use it to modify your career plan (see the chapter on career management). Once you have added this information to your career plan and decide that you still want a co-op experience, you can move forward with a clear understanding of what you want out of such an experience.

Next, take the information from your career plan and develop an executive summary of why you want a co-op experience and what outcomes you expect from that experience. This will be very important when you seek out companies to help you co-op with them.

Now to seek out companies that may offer you such an experience. We would start with the unit on your campus that deals with helping graduates and alumni find employment after graduation. There should be a list of companies that normally recruit on campus. These would be the most likely to consider offering you a co-op experience or an internship.

Another excellent source for candidate companies are the various advisory committees for departments, colleges, and the university. Typically they are composed of representatives that recruit and hire graduates, and the people on those committees would be the ones most likely to help you secure a co-op experience or internship.

Once you have secured an appointment as a co-op or as an intern, you will again need to review your career plan and make changes where appropriate (see the chapter on career management). Pay particular attention to outcomes you expect from the experience. In addition, make sure to document your experience, especially as it relates to expected outcomes, in a manner that enhances your portfolio (see the chapter on information capture and portfolios).

At the end of each co-op period, document your feelings about the company and list the pros and cons of working with that company once you graduate. Co-op students often receive a job offer from the company with whom they have worked. Make sure you record your impressions right after completing that period, since our memories tend to fade.

Also at the end of that period, determine whether you want to try for a new company or return to the same company for the next period. Your specific co-op requirements, however, may not allow you to return to the same company, so consult your school's requirements, if they have them, to find out what is allowed.

15.6 How to Get the Most Out of a Co-Op Assignment

Once you are at a company and participating in a co-op experience, make sure you do some very important things. First, you are hopefully very excited about what you have decided to do. You also need to make sure you have modified your career plan to reflect your co-op/internship experience. As part of this planning you should have already decided what information you need to gather for your portfolio as well as any reports you need to prepare for the company and your school. Work on these things as you go and do not wait until the last minute to do them.

> Your co-op experience is like almost everything you do, you need to focus on not only getting the most out of the experience but also make sure you are doing all you can to make your colleagues and supervisor look good. This is a great habit to have no matter where you are in your career.

15.7 Conclusion

There are many clear benefits to becoming involved in a co-op program or having internships while attending school. Briefly those benefits are additional skill development and enhancement, a source for self-motivation, a better perspective as to what an engineer is, opportunities for employment, and a good source for additional income during the process of obtaining an undergraduate or a master's degree. The only real downside to doing a co-op is that it will take you longer to earn your degree. This then makes the decision to participate fairly straight forward but maybe not an easy decision to make. Your career plan should prove to be very helpful to you in making this decision.

If you have decided to participate in such an activity, then you must make sure you have thoroughly integrated it into your career plan and have clear, stated expected outcomes from the experience. Make sure you fully document your activities by writing required reports and placing additional material into your portfolio.

END OF CHAPTER REVIEW QUIZ

Select the most appropriate answer to the following statements.

1. As a student your goal should be to
   - **a.** Become a professional student.
   - **b.** Be the best student you can be.
   - **c.** Prepare yourself to be the best possible engineer you can be.
   - **d.** Obtain the highest paying job you can get.

2. A co-op experience will result in
   a. Better skill development and enhancement.
   b. Higher starting salaries.
   c. An enhanced resume.
   d. All of the above.

3. Another advantage of the co-op experience is
   a. Better grades in the classroom.
   b. Better communication with your instructors.
   c. Better relations with your fellow students.
   d. Better understanding of the *"fun factor"* in an engineering career.

4. The benefits of a co-op experience are
   a. Skill development and enhancement and increased motivation.
   b. A better understand of the field of engineering and increased opportunities for employment.
   c. A good source for outside funding while earning your undergraduate and graduate degrees.
   d. All of the above.

5. Before you seek out a co-op experience you must make sure that
   a. Your school has a co-op program.
   b. A co-op experience is part of your career plan.
   c. You need the extra money.
   d. You find a good company with which you have the co-op experience.

6. Before you accept a co-op assignment you should
   a. Learn as much as you can about your school's co-op program.
   b. Go to the Internet and learn as much as you can about the company or companies you are considering.
   c. Interview students who are or have been co-opting with the company or companies you are considering.
   d. Do all of the above.

7. You should make your decision to participate in a co-op experience based on
   a. Your career plan.
   b. Advice from your advisor.
   c. Advice from other students.
   d. Advice from the dean's office.

8. At a school that has a weak or nonexistent co-op program, the best way to find companies that might be worth considering is to
   a. Seek out members of the advisory boards of the college and/or the department.
   b. Network with members of your local engineering organizations.
   c. Seek out students who are currently working and going to school part time.
   d. Do all of the above.

9. Relative to your career plan, you should
   a. Document your co-op experience carefully and completely.
   b. Make modifications based on your experience.
   c. Make sure your plan reflects whether or not to have another co-op experience in the future.
   d. Do all of the above.

10. One of the most important things to keep in mind while creating the best co-op experience as well as later in your career is to

a. Make sure you make the most money.
b. Make your colleagues and supervisor look good.
c. Dress appropriately.
d. Show up on time all the time.

EXERCISES TO DEVELOP AND ENHANCE YOUR SKILL SET

Exercise 15.1 To assess whether you want to consider a co-op or internship experience, start by writing down what you see as advantages to you of doing a co-op or internship. Now write down what you see as negatives of the co-op or internship experience. At this point, all you need to do is to determine if you want to proceed further in investigating the experiences available to you. If your negatives seriously outweigh the positives, then you can decide not to go any further at this time.

Exercise 15.2 Does your current school have a required co-op program? Does it not have a required co-op program, but does have a strong voluntary co-op program? Has your school almost no support for a co-op program? Once you have answered these questions, outline the details of the program or lack of program that you have.

Exercise 15.3 Finally, if you do go forward with seeking a co-op or internship, develop your plan for getting the most out of your experience.

CHAPTER

16 Graduate School, Maybe the Ultimate Way to Enhance Career Success!

"Education is the movement from darkness to light."

—Allan Bloom

"Wait a minute; I thought I only needed a bachelor's degree in engineering?"

When is the best time to think about going to graduate school? Right now is the best time to think about going to graduate school! In the chapter on career planning, we emphasize that your career starts in the first year of college if not before. Clearly, you need to understand whether graduate school is something you should begin planning for or to realize that it will play no role in your career. In this chapter, we answer questions such as whether you should earn a master's degree, whether you should pursue an MBA, whether you should go to the school you are currently attending, if it would be better to wait until you have some industrial experience, and whether you should earn a doctoral degree.

16.1 Introduction

Clearly, the twenty-first century belongs to the engineer. Every country that wants to have a high-quality standard of living needs a large and effective engineering workforce. A country's engineering workforce is the most valuable resource it can have. This need will significantly increase the demand for engineering graduates at all levels. We are indeed fortunate to be working during a time when engineering can have such a positive impact on our collective future!

You will indeed be entering the workplace during the most exciting time in engineering history. This chapter will help you to plan how your education will enhance your career. We explore the many myths about graduate education. As clearly as possible, we explain the benefits of graduate education and how it will make your life better. We look at issues such as whether you should earn a master's degree in business administration, whether you should you be a full-time or a part-time student, whether you should go to graduate school where you earned your undergraduate degree, and whether you should you pursue a doctoral degree. We address all of these issues and more.

LEARNING OBJECTIVES

By using the information and exercises in this chapter you will be able to

- Fully understand what an undergraduate degree in engineering is designed to prepare you to do in a career in engineering.
- Understand what a master's degree prepares you to do.
- Understand what a doctoral degree prepares you to do.
- Understand the value to your personal career of earning a master's degree.
- Understand what a doctoral degree will contribute to your career so that you can decide whether or not you should earn one.
- Know when is the best time for you to start graduate school.
- Know where you will attend graduate school.
- Find out if you should earn an MBA.

16.2 What Is a Bachelor's Degree in Engineering?

The best way to understand the role academia can play in your career requires that you have a clear understanding of the goals and expected outcomes of each degree. Let us begin by stating that each degree in engineering represents certain skills the student acquires.

Simply, education at the bachelor's level attempts to impart an understanding of the "language of engineering" and an ability to "think"; at the master's level, the student should acquire an ability to do advanced engineering and to communicate that work effectively both orally and in writing; the doctoral degree should represent a thorough understanding of the fundamentals of electrical engineering, and a mastery of the skills necessary both for working at the frontiers of almost any engineering area and for communicating one's efforts to others.

Many schools have tried to add more and more material to the undergraduate program in an effort to "more fully educate" the undergraduate. All such efforts have failed because those schools have missed the point of engineering education. Engineering education is much like teaching students to play football or any other organized sport. You can rush the acquisition of information but it is virtually impossible to speed up the process of developing and enhancing the skills necessary for a successful career in engineering.

> Another way to look at education is that poor education gives the student information that will become obsolete shortly after graduation, while coaching students to help them develop and enhance the skills necessary for a successful career in engineering will never result in the engineer ever becoming obsolete as long as they keep using their skills.

> ***"Give a man a fish and you feed him for a day. Teach a man to fish and you feed him for a lifetime."***
>
> –Chinese Proverb

What we really can do at the undergraduate level and do well is to teach the undergraduate an understanding of the "language of engineering" (additionally, the student will learn one or more dialects such as chemical engineering, civil engineering, electrical engineering, and mechanical engineering to name a few). This is followed by learning some subdialects such as thermodynamics, communication theory, computer design, structures, biomedical engineering, and the like. Another extremely important skill you will learn at the undergraduate level is the ability to "think like an engineer." Many seniors do not understand this, so they fear graduation because they feel they "really have not learned anything." They do not realize that industry knows this and will guide the recent graduate into an environment where engineering skills can be developed and enhanced.

One last thing—some students seek to earn a dual degree. They understand the value of having bachelor's degrees in, for example, electrical engineering and in mechanical engineering. This actually does not work as well as it sounds. If you truly want a dual degree, earn a bachelor's degree in one engineering area and then earn a master's degree in the other area. Not only will this result in much higher pay, it will also require fewer hours. Most master's programs require thirty semester hours (or forty-five quarter hours). If you do not have a bachelor's degree in the same area but have earned a bachelor's degree in another engineering area, you will have to take a few bridge courses to make up for missing prerequisites. Knowing that you want to do this, you can take the bridge courses as electives in your senior year.

> If you want a *dual* degree, make it a bachelor's degree in one engineering area and a master's degree in the other area!

16.3 What Is a Master's Degree in Engineering?

Of all the degrees we can earn in engineering, the master's degree is probably the most misunderstood. The recent graduate does not have to go into industry to obtain engineering skills, but may instead go into graduate school and acquire engineering skills by working toward advanced degrees. At the master's level the student will do some specialization in a given area and will learn to do advance engineering in that area. The student will tend to concentrate coursework in an area that supports a general interest. For instance, one may specialize at the master's level in topics related to software engineering and then write a thesis representative of an advanced engineering report in that area.

A master's degree

- Gives you the most employment opportunities of any degree.
- Allows you to do advanced engineering from the beginning.
- Allows you to manage engineering projects.
- Allows you to actually do design and to work at an advanced engineering level.
- In many degree areas, allows you receive a significant salary increase over the bachelor's level.
- Results in significantly enhanced communication skills.

Students ask us "What should I do if as a senior I do not know what to study in my master's degree program?" We reply that most of us do not know, in our senior year, in what area we wish to specialize. That is another advantage of attending graduate school and earning a master's degree! You will find that graduate courses will allow a clearer focus on specific areas. The student should be highly competent in one or more areas by graduation. Many graduate courses are developed over basically the same concepts that an undergraduate may have studied. The student now has more time to probe more deeply into these topics, because, in many cases, it is the second time for the study of those subjects. In general, it is easier because fewer, more interrelated courses are taken and the student's skill level has increased. This last item is something that students need to understand. As you develop and enhance skills, courses that might normally seem more difficult actually become easier.

Many students ask us if they should take a nonthesis option at the master's level. We have only one answer for that question. No! Aside from engineering skills, a student must also develop communication skills. We have found that, in general, success as an engineer will be directly proportional to the engineer's ability to communicate both in the written and oral forms. We also tell the student, "If you have problems in these areas please don't hesitate to take additional composition or speech courses." Writing a master's thesis is a must. One of the major skills acquired in the master's degree program is the ability to write an advanced engineering report, which should be a comprehensive and easy to understand document. The student is normally required to orally defend that thesis in front of a group of faculty members. So, in summary, the skills the student should acquire while in the master's program are an ability to work at an advanced engineering level and an increased ability to communicate.

Your success as an engineer will be directly proportional to your ability to communicate in both the written form and the oral form.

One of the myths of advanced education is that it limits your employment opportunities. Actually the greatest employment opportunities belongs to the graduate who has earned a master's degree in an engineering subject. If you look at the employment opportunities for a graduate with a bachelor's degree you will find that they can be divided into engineering jobs and nonengineering jobs, see Figure 16.1. It seems

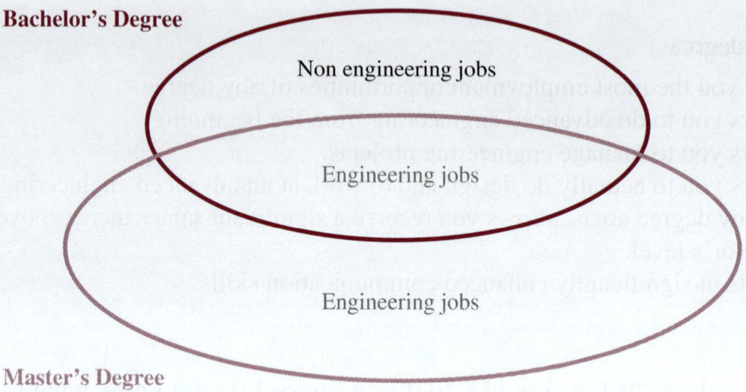

Figure 16.1
Employment opportunities.

that many employers like graduates with engineering degrees, which explains why so many engineering graduates are recruited for nonengineering jobs. Another interesting observation is that the engineering jobs that may be offered to graduates with bachelor's degrees typically will give preference to graduates with master's degrees. In addition, graduates with master's degrees are highly sought after.

To see how much employers value the master's degree just look at starting salaries. In some areas, a master's degree will give you an additional $15,000 per year. This is an excellent return on investment for one or two years beyond the bachelor's level.

A question we are frequently asked by students is, "Should I earn an MBA?" We have given this a great deal of thought and have only been able to arrive at one answer. A student considering earning either a master's degree in engineering or an MBA program should choose the master's degree in engineering! We have never seen the MBA be more beneficial to an electrical engineer, and in many cases the master's degree in engineering is clearly, more beneficial. Most engineers who enter management want to manage in an area related to engineering, and are much better off with the skills that are acquired in the master's degree program, especially, when it comes to supervising engineers. A person with a bachelor's degree in engineering followed by an MBA is normally not qualified to manage engineering operations. Obviously, there are exceptions to this rule, especially when one might have a great deal of engineering experience before entering the MBA program, but in general, the master's in engineering is definitely preferable.

Should I earn an MBA right after graduation with a bachelor's degree?
In general, "NO!"

We are frequently asked, "Who should enter a master's degree program?" Again, after a great deal of consideration and drawing from a long list of experiences, the only answer we can give is anyone who can, *must* obtain a master's degree in engineering. We would say the only qualification, which is important, is the word "can." An individual who has the desire and the ability to earn the degree should obtain it. There are no logical reasons for not earning such a degree. Even if the engineer does not end up practicing in the engineering profession for a complete career, the skills learned still make the master's degree in engineering very desirable.

> "Who should enter a master's degree program?"
> "Anyone who can, *must* obtain a master's degree in engineering."

16.4 What Is a Doctoral Degree in Engineering?

Like the master's degree in engineering, there are many things about the doctoral degree (this section refers to both PhD degrees as well as Dr.Eng. degrees) in engineering program that are misunderstood. The first item is that the doctoral degree in engineering is actually a very broad-based degree program. Candidates are normally required to understand at least four fundamental areas very well, well enough to be qualified to teach in any of those areas at the undergraduate level. The doctoral candidate should develop the ability to "finally understand those basic courses taken as an undergraduate." For instance, the student may eventually specialize in antennas and could be responsible for knowing the basics of mathematics that an engineer uses, the physics fundamental to a number of areas in electrical engineering, the basics of electromagnetic fields and waves, and digital system design and networks.

> A doctoral degree in engineering provides
> - An understanding of the fundamentals of mathematics, physics, and of engineering.
> - An ability to go to the state of the art in an engineering area and to move it forward.
> - An ability to communicate (writing, reading, speaking, and listening) at the highest possible level.

This broad-based educational process is important because it is part of the skill level that is developed by the doctoral in engineering candidate.

The most important skill (other than communication skills) the doctoral candidate acquires is the ability to go to the state of the art in almost any engineering field. Once at the state of the art, the techniques necessary to make an original contribution must be acquired. Finally, communication skills (writing, reading, speaking, and listening) must be significantly improved.

We are asked, "Who should earn a doctoral degree in engineering?" This is actually fairly easy to answer. A student who wants to do research, to start a business, to work at the frontiers of engineering, or to teach in a university, *must* earn a doctoral degree in engineering.The doctoral degree in engineering requires a great deal of effort and concentration. A decision has to be made by the student as to whether or not to remain a student for the two to four or more additional years required by such a program.

Note that the doctoral degree in engineering does not necessarily limit an engineer to any one particular area. The doctoral degree in engineering program emphasizes the development of skills, the skills to go to the state of the art in any given area, and then to contribute originally in that area. What a student chooses as a dissertation topic is of secondary importance to the skills obtained. For instance, a dissertation may be written in the area of stochastic controls; the new doctor in engineering graduate may then take a new direction and go quickly to the state of the art in antenna design, making contributions there, followed by the same in the area of digital system design.

One last thing about the doctoral degree in engineering is that approximately 83% of those with doctoral degrees in engineering go into industry and not academia. This is something that is normally found in only engineering doctoral programs.

16.5 When Should You Go to Graduate School?

Relative to all of this, we are asked, "When should I go to graduate school?" This is a challenging decision for most of us, and the best advice we can give is that the decision of when to go to graduate school is a question of maturity and finances, maturity being the major concern and finances, we think, being the minor consideration. A student who is academically mature enough to handle a graduate program should go right away; the sooner the better.

Many students make a serious mistake when they decide to put off going to graduate school. Why? It is serious because a student is less likely to come back and earn a graduate degree after leaving the campus, and the additional experience of working in industry normally does not significantly help future employment. This last item is important. Many students have the impression that five or ten year's additional experience at the bachelor's degree level followed by a master's degree in engineering or a doctoral degree in engineering is preferable. Unfortunately, this is generally not true. Most of us would be better off earning our graduate degrees as soon as possible.

Let us go back to the value of experience in industry, "Is it valuable?" If the experience was meaningful and at the right level the answer would probably be yes. Unfortunately, with a bachelor's degree in engineering, the ten years experience that could be acquired will be at a different level. So a person who works ten years in industry, and comes back and earns a master's degree in engineering will actually lag the engineer who earned a master's degree in engineering right after the bachelor's degree in engineering in both salary and work assignments.

These differences are even more dramatic at the doctoral degree in engineering level. An engineer who puts off earning their doctoral degree in engineering and acquires twenty years of experience even at the master's of engineering level will find themselves twenty years behind the engineer who earned a doctoral degree in engineering without going out and obtaining experience in terms of salary, promotions, and work assignments.

Another serious problem is part-time education. If there is no other way to earn a graduate degree, attending school part-time while working for a company is worthwhile, but it may not be the most effective way to attain one's academic goals. The best way is to become a full-time student. An assistantship, for those eligible, is helpful, but the major effort should be devoted to education. Unfortunately, a full-time job may be the focus of attention, leaving little energy for taking courses at night or during the day, 3 credit hours, and perhaps 6 credit hours per term. Everyone to whom we have spoken who has worked on a degree on a part-time basis tells us that it was a mistake to do it that way. It would have been much better to have earned it right away, and, unless destitute, a student should be able to survive on loans and assistantships. Believe us, the sacrifices that the student makes initially will be offset significantly by the job satisfaction that will accrue to the student who has earned advanced degrees as quickly as possible.

16.6 Where Should You Go to Graduate School?

Other frequently asked questions are "What school should I attend? Should I earn my degrees from three different schools or all from the same school?" At one time, when schools were seriously inbred, it was unwise to earn all three degrees from the same school. Today, however, earning degrees from the same school generally should not have any negative effect on employment and accomplishments, and, in fact, may be desirable since a student should be able to get through the program more quickly, developing and enhancing the student's skills. If your school is cosmopolitan with a number of faculty members from different universities, you can earn all three degrees from the same school without impacting your employability. Fortunately the vast majority of engineering schools fit into this category.

What advantages are there to earning all degrees from the same school? A student who is well respected as an undergraduate will probably be able to coordinate an entire graduate program related to previous undergraduate activities. Also, the faculty knows the student. The student may be able to save an academic term or more in the total time it takes to earn a degree. Some schools have accelerated master's programs, which can save as much as 12 hours out of the normally required 30 semester hours (reducing the requirements by 40%). On the other hand, if the student is having problems at a school and has developed a negative relationship with the department, then my advice is to go elsewhere. We like to say, "If you have fooled the faculty into thinking you are great, stay, if you have convinced them that you are not very good, it is a good idea to seek out another school."

If you decide to attend another school, do your homework! Earning a valuable graduate degree requires that you understand an important thing related to the educational process. You will learn as much outside the classroom as you will in the classroom, maybe more. This means your school should have an environment where there is a lot of outside the classroom interactions going on. These interactions take place with other students and with the faculty. It is a good idea to visit the possible schools you might attend and interview some of the faculty and some of the students. You need to find a place where you are comfortable and challenged.

Let me summarize and add one additional point for graduate education. Graduate students should be active in their student professional organization (such as ASCE, ASME, AIChE, and IEEE). It will help professional development and will increase the value of degrees. Two key advantages to being active in your professional organization as both a student and a working engineer, is that you build up a network that will help you professionally with new jobs, promotions, and salary and technically with quick access to the best technical information.

> **Engineers should be constantly educating themselves, formally and informally, taking advantage of all means for education.**

Students should obtain a thesis-based master's degree in engineering. A person who wants to do research, start a business, teach, or be at the frontiers of engineering, should obtain a doctoral degree in engineering. Many of these candidates will be our outstanding teachers in the future.

16.7 Conclusion

The primary goals for a bachelor's degree in engineering are to (1) develop and enhance the student's facility with a language, one or more of its dialects, and in several of its su-dialects and (2) develop and enhance the ability to "think" like an engineer. The primary goals for a master's degree in engineering are to (1) develop and enhance a student's ability to do design and advanced engineering and (2) develop and enhance a student's communication skills. The doctoral degree in engineering seeks to (1) develop and enhance a student's ability to do research and state of the art engineering; (2) develop and enhance the student's understanding of the basics of mathematics, science, and engineering; and (3) develop and enhance a student's communication skills at the highest level possible.

Everyone who can earn a master's degree in engineering must do so. Everyone who wants to do research, to start a business, to work at the frontiers of engineering and technology, or to teach in a university, *must* earn a doctoral degree in engineering.

Earning a graduate degree may be the single most important asset in the development of a successful career. If you do decide to go to graduate school you should go right away, go full time, and go to the school where you earned your undergraduate degree. You need to make sure you take advantage of all the opportunities for skill development and enhancement outside of the classroom as well as inside the classroom.

Finally, earning an MBA will not enhance your career like earning a master's degree in engineering. The only time an MBA might be important would be once you have at least ten years of experience and have decided to go into management.

END OF CHAPTER REVIEW QUIZ

Select the most appropriate answer to the following statements.

1. The purpose of an undergraduate degree is to

a. Make you an engineer.

b. Help you to develop the skills so you can learn how to become an engineer.

c. Prepare you for graduate school.

d. Develop the knowledge you will need to become an engineer.

2. During your undergraduate experience you will learn
   a. A language associated with your declared major.
   b. How to really *think*.
   c. How to learn how to become an engineer.
   d. All of the above.
3. The purpose of a master's degree is to
   a. Turn you into an engineer and teach you how to communicate at a higher level.
   b. Make you an expert in an area in your declared major.
   c. Allow you to qualify for higher level engineering jobs.
   d. Decrease the extent of your career opportunities.
4. Who should earn a master's degree?
   a. Those who want to work in more restricted areas
   b. Anyone who can
   c. Those who do not know what they want to do
   d. Those who know what they want to do
5. When should you earn a master's degree?
   a. As soon as you can
   b. After you have had five to ten years of experience
   c. After you have earned an MBA
   d. Never
6. The purpose of a doctoral degree is to
   a. Teach you how to communicate at the highest possible level.
   b. Teach you how to do research and state of the art engineering in almost any engineering area.
   c. Help you learn the fundamentals of any engineering area.
   d. Accomplish all of the above.
7. Who should earn a doctoral degree?
   a. Anyone who wishes to do research and state of the art engineering
   b. Anyone who wishes to teach in a university
   c. Anyone who wants to start their own business
   d. All of the above
8. When should you seek a doctoral degree?
   a. As soon as you can
   b. After earning a master's degree and then having five to ten years of experience
   c. Only after earning an MBA
   d. All of the above
9. Where should you earn your graduate degrees?
   a. If you have shown the faculty at your school that you are a good student and you like the environment, stay at the school where you earned your undergraduate degree.
   b. To enhance your marketability you need to earn your graduate degrees at different schools.
   c. You should earn your degree at the school that has the best reputation.
   d. None of the above applies.
10. Which degree provides access to the largest number of engineering jobs?
   a. The bachelor's degree
   b. The master's degree
   c. The MBA
   d. The doctoral degree

EXERCISES TO DEVELOP AND ENHANCE YOUR SKILL SET

Exercise 16.1 Since you really must obtain a master's degree if at all possible, you need to assess whether or not it is possible. You may not have the grades to be accepted into a graduate program, then you need to ask yourself, do you think you might ever want to pursue a master's degree. If you do not feel like the master's degree is in your future, then you need to focus on your career and develop a plan for skill development and enhancement throughout your career. If you do decide that you will most likely pursue a master's degree in the future, then develop a plan to make up the deficiencies you and your advisor identify. In many cases, there is a way to earn this degree in spite of not having good grades. See your advisor for help with this.

Exercise 16.2 Identify the schools you wish to consider for pursuing a master's degree. Write down all the pros and cons relative to each school. Make sure you actually visit any school you are seriously considering. Once you have done this, make a decision on which of the schools to which you wish to apply (it is always a good idea to apply to at least two schools).

Exercise 16.3 For those of you who may or may not wish to pursue a doctoral degree, develop a plan for obtaining the degree. The first thing you should do is to develop a list of pros and cons about pursuing the degree. Your advisor will be an asset in this entire process. Next, you need to develop a plan to pursue the degree. It is also advisable to develop a plan in the event you do not pursue the degree. Review both plans along with the pros and cons to eventually make a decision about pursuing the degree. Your plan to pursue the degree should also include choosing the appropriate schools to apply to, as you did in Exercise 16.2.

Reference

"A Solution to the Crisis in Engineering Education?" Charles Alexander, *IEEE Transactions on Education*, May, 1984.

CHAPTER

17 Through the Looking Glass, Bring All the Skills You Have Acquired to Build a Successful Career

"I wonder if I've been changed in the night? Let me think. Was I the same when I got up this morning? I almost think I can remember feeling a little different. But if I'm not the same, the next question is 'Who in the world am I?' Ah, that's the great puzzle!"

—Alice

Now that you have covered all of what we want you to understand about creating and having a successful career, you need to address how to make the transition between being a student to being a practicing professional. In some respects it is like Alice (from Lewis Carroll's, Through the Looking-Glass*), when she "stepped through the looking glass!" I am sure she, like you, was terrified and really did not know what to expect. You need not be afraid; you are so much better prepared for this step than most of your colleagues that you cannot help but to be successful. In this chapter, we will relate what you have already learned with what will be expected from you in whatever assignment you accept. We will help you with your career planning and professional development as a working engineer.*

17.1 Introduction

Instead of acquiring a knowledge-based degree, you have trained yourself just like an athlete. You have become highly skilled in all the tools necessary for a successful career in engineering. Since transitioning from being a student to the role of practicing professional is such a major transition, you need to develop the confidence that you can be successful.

Fortunately almost every new engineer finds this transition scary at first but once the transition has been made, they are incredibly surprised how easy it actually was. The most common response is that they are now having fun as a practicing professional!

LEARNING OBJECTIVES

By using the information and exercises in this chapter you will be able to

- Know if you really want to leave school.
- Know that you really are ready for the "real world."
- Better understand the importance networking to your career.
- Better understand the transition between being a student and a practitioner.
- Better understand the need for constant skill enhancement.
- Know how well you are prepared to be successful.
- Know that you may never really retire.
- Know that you really will be okay.

Why fun? The answer is simple, as in the example of the trained athlete, once you get very good at what you do, you enjoy what you do that much more. In addition, as a student everyone controls you, as a practicing professional you are now in control and that feeling is indescribable!

17.2 Do I Really Want to be a Professional Student?

It is very tempting to want to stay in a place where you feel comfortable. Many students think that staying where they are and not changing where they are is an okay decision. First, you need to make sure you have the highest degree you feel you need. Once you have that, then you need to break your comfort level and "get out into the real world!"

Please understand the difference between being a professional student and engaging in lifelong learning. It is the latter that we all must do if we are to have a long and successful career. The former addresses those students who just randomly take courses to avoid graduating and securing a position in engineering or related areas. Fortunately this applies to very few students in engineering; most of us could not wait until we got out and started doing things! For those students who see themselves drifting toward the professional student status, do everything you can to control these feelings; you will never regret making the transition.

Again, we can use the example of an athlete, practicing and never playing in a real game means you will never know the thrill of performing competitively. Get out and play!

17.3 Stepping Through the Looking Glass, Good or Evil?

Stepping through the looking glass is definitely good not evil! Just think of how many times you wish you could design something that actually worked! Many of you will successfully complete a senior design project and obtain a taste of what the real world is like. For those of you who have co-op experience will also have a taste. Let us share a personal experience with you:

> Before we graduated we had some idea what it would be like to go out into the real world. However, our actual experience is that what really happened was so much more exciting that anything we expected!
> "ENGINEERING IS FUN!"

Note especially that you need to continually follow the skill enhancement guidelines we have presented in the other chapters of this book and you will be successful. Remember, these have all been presented as the skills you need for career success.

> The successful practicing professional will constantly enhance their skill set. A significant advantage to this is that the professional who follows this advice will continue to become better and better no matter what their age. We have seen firsthand that these skills rather than knowledge allows for constant improvement.

The best measure of this is that for engineers with just the bachelor's degree, their salaries continue to rise until their early to midfifties when their salaries start to decline but not in a major way. With a master's degree salaries continue to rise with a leveling during the early to midfifties. For engineers with their doctorates, their salaries continue to rise throughout their career. And, because engineering is a skill-based career, and engineering is fun, most of us continue to work into our seventies and eighties. And we do this not because we have to, we do it because we want to!

17.4 Making the Transition Is Easier Than You Think

How do you pick the best job for your first job? You need to apply the process discussed in the chapter on career planning along with the guidelines in the chapter on portfolio development to help you identify and secure your first or second choice for your first job.

We note that your first job may be the last job you seek using conventional job searches. If you develop your network properly, all future jobs, promotions, and raises will most likely come through your network. Please see the chapter on networking to pursue this in more detail.

Let us focus on securing your first job. The important thing to remember is that this is your first job, not your career. It may develop into your career, but most likely it will be just the first of many jobs you may have in your career. Therefore, a lot of the pressure is off your needing to make the perfect decision. Our advice is to even apply for positions that you may at first think you do not want. Sometimes your best choice may come from an unexpected opportunity. If you are fortunate and receive one or more job offers, then you need to determine whether or not to accept the job. Before you apply for a job, you should make a list of wants and a list of do not wants. Then, using your wants and do not wants as a basis, make a list of pros and cons relative to your offer. Doing this will lead you to making a more informed decision.

> The important thing to remember is that this is your first job, not your career. It may develop into your career but most likely it will be just the first of many jobs you may have in your career.

Once you have made a decision, you need to make sure that as you perform the duties of this new job you remember that the most important thing for your career is to make sure you continue to develop and enhance your skills as you have learned to develop and enhance them in this text. Another very important thing to remember is that you need to focus on doing things that will make your colleagues and supervisor look good. That may seem a little strange but we are all human beings and that regardless of how well you can do your job, promotions, new positions, job satisfaction, and raises will depend a lot on the good will you have created.

What do you do in periods when you do not have enough to do to fill your work week? Many of you will read the company publications to obtain a better understanding of your employer. Some of you will go into the lab and learn how to use many of the instruments that are in the lab. One thing we have found is to first go to your supervisor and make sure that it is okay for you to volunteer to work in another area for a while. Most of the time, your actions will be appreciated and can result in additional skill development and enhancement.

17.5 Conclusion

> ***"Your Happiness is Critical, pay attention to it daily! In the end, it's not the years in your life that count. It's the life in your years."***
>
> —Abraham Lincoln

One of the important decisions you will make is what you want to do about retirement. You might wonder why as a young person you need to think about retirement so soon after graduation. The answer is simple, decisions made now, especially financial ones, will have a far-reaching impact on your retirement.

Financial investments made long before you retire will grow substantially by the time you reach retirement age. We also want you to know that if you plan on traveling when you retire, make sure you travel before you retire so you are motivated and comfortable with traveling when you do retire.

We have seen folks who said they would travel upon retirement not do so because they have not done it before. Another thing, engineers are creative and so they are never happy just retiring. For most of us, retirement means finding another challenge we need to meet. We start new businesses, become consultants, and some of us become writers!

We would like to end on this very important thought:

> *"You are entering the most exciting period in the history of the world and the most exciting time to be an engineer!"*

END OF CHAPTER REVIEW QUIZ

Select the most appropriate answer to the following statements.

1. What is true about the education you have received?
- **a.** You will become obsolete in two to three years.
- **b.** You have the skills that, if maintained and exercised, will never become obsolete.
- **c.** You have learned all you can learn, now you need to *"practice."*
- **d.** You now must begin the lifelong struggle to keep up.

2. What is the best model to use to ensure a long and productive career?
- **a.** An accomplished academic with the curious mind to grow and prosper
- **b.** An exceptionally strong knowledge base
- **c.** Skills like those of an athlete that when developed and enhanced will lead to a long and successful career
- **d.** All of the above

3. If properly developed and enhanced, your skills will ensure that your career will
- **a.** Be successful at a very high level until your late forties.
- **b.** Make you highly desirable for at least fifteen years.
- **c.** Have the potential to allow you to increase your productivity throughout your entire career.
- **d.** Allow you to be successful if your youth and energy will support you.

4. Almost all of the engineers who have been successful would say what best describes their success is
- **a.** Fun.
- **b.** Hard work.
- **c.** Fire in the gut.
- **d.** Dedication.

5. How important is the decision you make to secure your first job?
- **a.** Your first job is not your career.
- **b.** Your first job is really only one part of your career.
- **c.** You need to perform in this first job at the highest possible level.
- **d.** All of the above are true.

6. What is the most important thing to keep in mind with your first job?
- **a.** There is only job security.
- **b.** There is no job security, only career security.
- **c.** If you are not successful in your first job, you can kiss your career goodbye.
- **d.** There is nothing you can do to have career security.

7. One of the most important things to keep in mind when performing in your current job is that
- **a.** You need to make sure you do a great job.
- **b.** You need to make sure you distinguish yourself from those around you.
- **c.** You need to make sure you make your colleagues and supervisor look good.
- **d.** You need to keep your next assignment clearly in your focus.

8. You are doing a great job and begin to have excess time on your hands, you should
   a. See about going home early.
   b. Take time off to go to school.
   c. Volunteer to help other teams within your workplace.
   d. Relax and enjoy the experience.
9. A key element for a successful career is your happiness in what you are doing, so you need to
   a. Make sure you maintain and enhance your career plan.
   b. Make sure you seek out those assignments that will make you happy.
   c. You focus on maximizing your salary.
   d. Do all of the above.
10. What is the most important thing to keep in mind about retirement?
   a. Most engineers never retire, mostly you just switch career paths.
   b. You must make sure that you can maintain your current lifestyle with your retirement income.
   c. You need to make sure you cover your health needs.
   d. You need to develop a trust for your children.

EXERCISES TO DEVELOP AND ENHANCE YOUR SKILL SET

Exercise 17.1 Now that you have completed this chapter, revise your career plan to reflect what you have learned.

Exercise 17.2 If your career plan does not already include a plan to obtain your first job, develop a plan to "land" your first job.

Exercise 17.3 Evaluate your career plan to see if it reflects that you have focused appropriately on career security and not job security. If you see that your career plan needs to be modified then make the appropriate modifications.

References

What Color Is Your Parachute? 2011: A Practical Manual for Job-Hunters and Career-Changers, Richard N. Bolles, Ten Speed Press, 2010.
The Job-Hunter's Survival Guide: How to Find a Rewarding Job Even When "There Are No Jobs", Richard N. Bolles, Ten Speed Press, 2009.
How to Find a Job on LinkedIn, Facebook, Twitter, MySpace, and Other Social Networks, Brad Schepp, McGraw-Hill, 2009.
Knock 'em Dead 2011: The Ultimate Job Search Guide, Yate Martin, Adams Media, 2010.
Job-Hunting Online: A Guide to Job Listings, Message Boards, Research Sites, the UnderWeb, Counseling, Networking, Self-Assessment Tools, Niche Sites, Mark Emery Bolles and Richard N. Bolles, Ten Speed Press, 2008.

INDEX